The Scholarship Series in Biology

General Editor: W. H. Dowdeswell

Animal Taxonomy

THE SCHOLARSHIP SERIES IN BIOLOGY

Human Blood Groups and Inheritance
S. Lawler and L. J. Lawler

Woodland Ecology
E. G. Neal

New Concepts in Flowering Plant Taxonomy
J. Heslop Harrison

The Mechanism of Evolution
W. H. Dowdeswell

Heathland Ecology
C. P. Friedlander

The Organization of the Central Nervous System
C. V. Brewer

Plant Metabolism
G. A. Strafford

Comparative Physiology of Vertebrate Respiration
G. M. Hughes

Animal Body Fluids and their Regulation
A. P. M. Lockwood

Life on the Sea-shore
A. J. Southward

Chalkland Ecology
John Sankey

Ecology of Parasites
N. A. Croll

Physiological Aspects of Water and Plant Life
W. M. M. Baron

Biological Principles in Fermentation
J. G. Carr

Studies in Invertebrate Behaviour
S. M. Evans

Ecology of Refuse Tips
Arnold Darlington

Plant Growth
Michael Black and Jack Edelman

Animal Taxonomy
T. H. Savory

Animal Taxonomy

Theodore Savory, M.A., F.Z.S.

Heinemann Educational Books Ltd
London

Heinemann Educational Books Ltd

LONDON EDINBURGH MELBOURNE
TORONTO AUCKLAND JOHANNESBURG
SINGAPORE HONG KONG IBADAN NAIROBI

SBN 435 61800 8

First published 1970

Published by Heinemann Educational Books Ltd
48 Charles Street, London W1X 8AH
Printed in Great Britain by
Richard Clay (The Chaucer Press) Ltd, Bungay, Suffolk

Preface

In our zoology texts there are few topics that are treated in so formal, almost lifeless a manner as is the subject of the classification of animals. Often the reader is told that we owe our system to Carl Linnaeus, and is given a tabular classification of the animal kingdom, more or less detailed, according to the intentions of the writer. This is presented in a confident form, as if it had been settled many years ago, like the list of the Kings of England, with never a suggestion that the groupings are under constant review, changing as our knowledge grows, creating their own language, and always striving towards a state of stability which at present does not seem to lie in the foreseeable future.

One purpose of this book is to modify this point of view by attempting to present taxonomy as a lively study, even in its own way an exciting one, possessing and giving all the pleasures of intellectual exercise, and above all of real importance as the fundamental expression of the principles of systematic zoology.

I am grateful both to Mr. Hamish MacGibbon, my publisher, and to Mr. W. H. Dowdeswell, editor, for their help and encouragement while I have been writing this book.

The book itself is one result of seven incredibly happy years spent in teaching zoology in a school with a high reputation for science scholarship. There I met boys far cleverer than myself, boys clearly destined for University honours, and whom I could but try to guide, encourage, and hope to inspire; and I want to take this opportunity to acknowledge my deep sense of gratitude to all those who, for so long, have borne patiently with my stumbling phrases and who, in their cheerful discussions, have done so much to help me to marshal my own nebulous ideas. Occasionally it is admitted that in our Public Schools the boy gains something from his master; but less often is it realized and admitted how much the master may owe to his boys.

T.H.S.

Contents

1

Approach to Systematics

Scientists in general, and perhaps biologists in particular, often assert that science cannot be learnt from books, but that it must be studied in the laboratory, or in the woods, ponds and fields, or some other comfortless place, but never by the fireside, in the study chair.

Of course they are right, or partly right, for no amount of reading would enable a student, otherwise unpractised, to expose the fifth cranial nerve of a dogfish or to remove a nephridium from an earthworm. But it is not every zoologist who wishes to do such things as these, or is able to continue to do them, and the study chair is not to be neglected. It is a place where a zoologist may profitably sit and think, a place where he may conceive and write a book, and enjoy both the thinking and the writing.

Every scientist, whether his personal concern is with physical or biological matters, comes quickly to perceive that the facts with which he has to deal, and the hypotheses which they suggest, arrange themselves in a pattern or system, often of an unexpected nature. It seems as if there were a fundamental orderliness in all natural phenomena, from which some underlying principle could be deduced, and that all the facts discovered and all the theories created could be fitted into appropriate places in the philosophy of the science involved. As soon as the particular system has been revealed, progress in the science becomes more rapid and inspired with a greater confidence. In other words, every science has its own kind of systematics, so that, as the American zoologist G. G. Simpson has said, systematics is the most scientific of all the sciences. The words 'systematic' and scientific' should, he says, be interchangeable. If this is true, every zoologist should have an adequate knowledge of systematics.

We zoologists classify because we must. We find so many different animals in the world that we cannot treat them separately: even if we wanted to do so the task would be beyond the capacity of the human mind and memory. Classification is forced upon us by the limitations of the brain. There are some headmasters who know by sight every boy in their school, some colonels who know by sight every man in the regiment; but these are exceptional persons. For most of us a limit is set well below the thousand mark.

We also classify unintentionally, or, as some might say, instinctively, so that we seldom hear the question, 'why classify?'. From the earliest times our predecessors found it to be advisable to note and to act on the differences in the multitude of things that surround them. They distinguished men who were enemies to be slain from those who were friends to be encouraged, they discovered that some plants were nauseous to the taste but that others were edible, and they learnt that some animals were ferocious and were to be avoided while others were docile and could be domesticated.

If, however, there remains any wish to insist on an answer to a 'why' question, it may be found in a few more words written by Simpson, when he assured us that 'the primary purpose of classification is simply to provide a convenient practical means by which zoologists may know what they are talking about and others may find out'.

So there is no need to ask why animals are classified. Classifying is an innate mode of thinking, and leads to a study that reveals both differences and similarities between the objects concerned, whatever they may be.

Imagine, if you will, the blade of a cricket bat with an umbrella handle fitted into the splice. A hole in the blade is closed by a lens, made ineffective by a piece of shark skin held behind it by two hob nails. The bottom of the blade is wedged into a horseshoe, with a piece of ribbon threaded through its holes. Into which section of a Universal Stores catalogue would you place this nightmare object?

The purpose of this absurdity is to show that a unique object cannot be classified. Classification demands comparison with other, similar, entities, and is founded on a detection of the differences between them.

Try this experiment. On one side of a piece of paper draw two squares: colour one of them red and the other blue or green. On the other side draw a square and a circle: colour them both red, or blue, or green. Present either side to anyone and ask, 'What is your first reaction and your first comment on seeing these?' Then ask him to turn the paper over, and repeat your question. You will find that almost invariably the answers given are 'They are different colours' and 'They are different shapes'; and very seldom 'They are the same shape' and 'They are the same colour'.

We all tend, indeed, to notice and to pay more attention to differences, and this inclination has had a considerable effect on our attempts to classify animals in an acceptable way. Classification must be based on both similarities and differences, yet it always seems to be easier to detect and judge differences than to recognize and evaluate similarities. Consequences of this psychological bias will be seen again and again in any study of the principles of classification.

Clearly there must be some general agreement upon the principles by which our classification of animals is to be guided. The science of classification is called taxonomy. Since it has been shown that there is no profit in the question 'why classify?', it must be replaced by the alternative 'how do we classify?'.

Taxonomists, like other men, very soon discover that there are several ways in which any group of objects can be classified, and an important feature of this may be illustrated by considering an analogy. Books are more familiar to most of us than are many animals, and books can be classified on their shelves in various ways. They can be arranged according to their colour, their subject matter, their authors' names, their ages, and even their sizes. It is easy to imagine circumstances in which any one of these methods might be the most appropriate. But, it may be asked, what is really the best way?

The answer is that this depends on the reason for which the arrangement was undertaken, or on the purpose of the classifier. Think of the question, 'what is the best knife?'. Clearly it depends on the use to which the knife is to be put; and obviously the answer to the question 'how can we best classify animals?' is that it depends on the purpose of the classifier.

Fundamentally there are two different ways in which the animals of the world have been classified, the archetypal method and the hierarchic method.

In the former, each group of animals is composed of those that show common characteristic, such as the possession of six legs. Each such group contains within itself other groupings, similarly composed of the animals with other common characteristics, such as the possession of two wings, in addition, of course, to the feature used to determine the inclusive group.

A simple example, based on a few myriapods, insects and arachnids, will make this clear.

A Animals with bilaterally symmetrical bodies	
I Animals with legs	
1 Animals with ten or more legs	
a Animals with two legs per segment	Centipedes
b Animals with four legs per segment	Millipedes
2 Animals with fewer than ten legs	
a Animals with six legs	Insects
i Animals without wings	Apterygota
ii Animals with wings	Pterygota
α Animals with two wings	Diptera
β Animals with four wings	Lepidoptera
b Animals with eight legs	Arachnida
i Animals with terminal silk glands	Araneae
ii Animals with terminal venom glands	Scorpiones
II Animals without legs	
etc.	
B Animals with radially symmetrical bodies	
etc.	

In the above way of representing an archetypal classification the most noticeable feature is that, if fully developed, any two equivalent steps, such as *I* and *II* above, may be a long way apart. In a comprehensive classification several pages may separate them. This is avoided by recording the scheme in the form of a dichotomic table, usually called a key. In a key the alternatives are brought closer together, as is shown below. There are several different ways in which keys are printed: in the following key, numbers in parentheses are alternatives to the numbers that preceed them.

1 (2)	Animals with bilaterally symmetrical bodies	3
2 (1)	Animals with radially symmetrical bodies	18
3 (4)	Animals with legs	5
4 (3)	Animals without legs	17
5 (6)	Animals with ten or more legs	7
6 (5)	Animals with fewer than ten legs	9
7 (8)	Animals with two legs per segment	Centipedes
8 (7)	Animals with four legs per segment	Millipedes
9 (10)	Animals with six legs	11
10 (9)	Animals with eight legs	15
11 (12)	Animals without wings	Apterygota
12 (11)	Animals with wings	13
13 (14)	Animals with two wings	Diptera
14 (13)	Animals with four wings	Lepidoptera
15 (16)	Animals with terminal silk glands	Araneae
16 (15)	Animals with terminal venom glands	Scorpiones
17 (18)	Etc.	
18 (17)	Etc.	

Each pair of lines in the above divides the animals into two groups, hence its description as a dichotomic table. Keys of this kind have great practical value, and a well constructed key should enable anyone, even with no great knowledge of the animals concerned, to use it with success. For this reason they have been used for centuries by naturalists wishing to determine the names of the specimens that they have found. They are the chief components of most of the botanists' 'Floras', and are very often to be found in scientific papers which deal with the species of a genus or the genera of a family which their writers have monographed, as well as in accounts of the flora or fauna of a particular region, island, or county. But it may be added that unless they are carefully compiled they are apt to justify the jibe that one can always identify a strange object with the help of a key provided that one already knows the other specimens in it.

Attention should be called to the use of the word 'identification' in the above paragraph, for it implies the noticing of points of close similarity between the objects concerned. The imaginary artefact described earlier could not be found a suitable place in a general catalogue because it was like nothing else, although it could have been classified as a composite solid or perhaps even as a specimen of modern sculpture.

The second, the hierarchic method of classification is invariably used by zoologists. In a hierarchy there is no attempt to concentrate attention on pairs of contrasting characteristics. Each group, whether it be class, family, or any other, is defined by a certain number of features, the majority of which are shared by all its members. Each group or taxon, to use the most recent term, except the lowest, is divisible into a number of subsidiary groups, also with characteristics of their own. Each taxon is named. The small selection of animals chosen to illustrate the archetypal method would be expressed in a hierarchy as follows:

ARTHROPODA				
MYRIAPODA	INSECTA		ARACHNIDA	
	APTERYGOTA	PTERYGOTA	CAULOGASTRA	LATIGASTRA
CHILOPODA DIPLOPODA		DIPTERA LEPIDOPTERA	ARANEIDA	SCORPIONES

The construction of a chart of this kind is very informative but suffers from the disadvantage that if it is either comprehensive or detailed it requires an enormous and unwieldy piece of paper, which is also impossible of reproduction in ordinary printed books.

If it is compared with a similar arrangement of a human pedigree the resemblance is too obvious to escape notice.

<table>
<tr><td colspan="6">GRANDFATHER</td></tr>
<tr><td>UNCLE</td><td colspan="4">FATHER</td><td>AUNT</td></tr>
<tr><td>COUSIN</td><td>BROTHER</td><td colspan="2">SELF</td><td>SISTER</td><td>COUSIN</td></tr>
<tr><td></td><td>NIECE</td><td>SON</td><td>DAUGHTER</td><td>NEPHEW</td><td></td></tr>
</table>

This forces attention on the chief feature of the hierarchic system, which is that it aims at representing or reflecting the course of animal descent which we call organic evolution. When a list of classes or families of animals is printed, it is not printed alphabetically in the form of an index. We do not write 'Amphibia, Birds, Fishes, Mammals, Reptiles', but 'Fishes, Amphibians, Reptiles, Birds, Mammals', with the implication that in systematics phylogeny and taxonomy are to be united as closely as possible.

Phylogenty includes the imaginative side of theoretical zoology. Its subject is the evolution of animals and the object of phylogenists is the tracing of the courses that evolution has taken. It is therefore an uncovering of a buried past, and the unravelling of a tangled skein, and since in the burying much has been destroyed and in the tangling many threads have been snapped there are sufficient opportunities for inspired guessing.

The last word is used deliberately. Guesswork is not to be underestimated, and in this connection speculation is both legitimate and necessary. When speculation is rational it is likely to meet with corroboration, when it is irrational or mistaken it will sooner or later be forgotten.

Taxonomy is the study of the principles and methods of classification. A taxonomist has not only to detect the differences and the similarities between animals as he studies them; he has also to decide whether differences are due to anything more than adaptation to different environments, and whether the similarities are the result of a common ancestry or of convergent evolution. He must face, too, the difficult question of the weighing of characteristics: is a biramous limb more or less specialized than a uniramous one, and so on. If he makes the wrong choice his classification will not be truly phyogenetic. These and other features of taxonomy will find places in the chapters that follow.

It should also be pointed out that the hierarchic system is a classification based fundamentally on homology in organs and organisms, which can be grouped naturally by selecting modifications in homologous parts. If, on the contrary, a system is based on analogy or on analogous characters it can only take the form of a key, of which the following example illustrates an important feature.

This is that a key can be constructed to distinguish the members of a group of objects of any kind whatever: indeed the composition of such a key is a valuable excercise for the student who is making his first acquaintance with taxonomy. Thus:

1(2)	Longer than broad	3
2(1)	Broader than long	5
3(4)	Pointed at one end	Biro pen
4(3)	Pointed at neither end	Cigarette
5(6)	Solid	Apple
6(5)	Hollow	Tennis ball

As has been said, one purpose of systematics is to enable zoologists to know what they are talking about. But they cannot talk about animals until the animals have names by which to refer to them. Hence nomenclature enters the field, bringing its own rules designed to meet its own difficulties. One might, perhaps, believe that when once an animal had been discovered, described, and named the matter would be at an end, but this innocent hope has not been fulfilled. A parallel may be illuminating. One might think that the correct addressing of an envelope should be a sufficiently simple operation, but in practice the G.P.O. has found it desirable to distribute a pamphlet of instructions to correspondents. Onomatography, or the correct writing of animals' names, demands attention, even from those who are not interested in nomenclature as such.

Like phylogeny, nomenclature is therefore a concomitant necessity in systematics, which thus assumes a tripartite structure:

SYSTEMATICS		
PHYLOGENY	TAXONOMY	NOMENCLATURE

Systematics is the synthesis of scientific thinking in zoology, and is indeed the only way to save ourselves from zoological chaos.

2

The Rise of Taxonomy

There is no doubt that in the study of the theoretical aspect of any branch of science, a historical approach is the most satisfactory. It exposes the unfolding of thought and the growth of ideas, and it includes a proportion of human interest by its mentioning of the men to whom progress has been due. This is certainly true of systematic zoology.

Like the history of zoology in general, or indeed like the history of any other branch of science, that of taxonomy begins with Aristotle (384–322 B.C.). His contributions were few but definite. He realized the existence of different degrees of resemblance and differences between animals, and expressed these in definitions of what we should now call genera and species. In this he was following Plato, who had similarly distinguished between the genos and the eidos, but his words show clearly enough that, although he was thinking along lines that later led to the idea of a hierarchic classification, he did not develop this into a coherent scheme for all the animals that he knew.

Yet Aristotle had an enviable knowledge of comparative anatomy and of embryology, and had made many records of his observations in various places in his writings. From such scattered diagnoses enthusiastic admirers have constructed an Aristotelian classification which Aristotle, in fact, never wrote. Some may feel that this is a zoology *pour rire*, but it has a value in so far as it shows how the genius of Aristotle may be said to have founded the science of taxonomy.

Fragmentary and inchoate as it was, Aristotle's system was not superseded for more than two thousand years. After the long silence of the Dark Ages the foundations of scientific zoology were laid by one of the greatest geniuses the world has known, Conrad Gesner of Zurich (1516–65). In his immense *Historia Animalium* of 1548 he was content to follow the method of Aristotle, and so to earn himself the unmerited name of the German Pliny. No scholar has been more justly famed for the range of his learning than was Gesner, and he made a few contributions to taxonomy. It can be said of him, as of his contemporaries, that he suggested both variations and improvements in the system that he had inherited.

The first substantial advances were due to a botanist whom all zoologists are proud to honour, John Ray (1627–1708), a native of Notley, Essex and a Fellow of Trinity, Cambridge. He was a voluminous writer, and the work that chiefly interests systematists was his *Synopsis Methodica Animalium Quadrupedum et Serpentini Generis*, published in 1693. Other, less important, works on insects, birds and fishes were published after his death.

In the Synopsis Ray followed the same dichotomic method that he had already used in the classification of plants. Like Aristotle he began by dividing animals into those with blood and those apparently without blood, and the former into those with gills and those with lungs. By continuing in this way he was able to cover almost the whole of the animal kingdom, making use of such characteristics as the production of eggs or of living young, the possession of broad hooves or of narrow claws, the existence of two or of more incisor teeth, and so on. His system was not only a logical one; within the limits of the knowledge of his time it had also the advantage of being a practical method that was easy to use. In consequence it was welcomed, and it remained in popular favour until it was displaced by the work of Linnaeus.

Ray died two years before the birth at Raschult in Sweden of Carl Linnaeus, later Carolus von Linné (1707–78), the son of a clergyman and subsequently professor of botany at Uppsala. At Leydon in 1735 he published a small book of eleven folio pages, bearing the now famous title of *Systema Naturae*.

He began with the natural empire and split it into three kingdoms, those of animals, plants and minerals. He introduced the hierarchic system into the animal and plant kingdoms, his animal world being divided into four grades—class, order, genus and species. Twelve editions of the book were published before his death, and of these Linnaeus himself revised the second, sixth and tenth.

Biology's chief debt to Linnaeus lies in his use of a binominal nomenclature for all species of plants and animals. The tenth edition of 1758, in which the name Mammalis was first used in place of Quadrupeda, was the first in which this nomenclature was consistently used throughout.

The significance of Linnaeus' achievement lies in the fact that he was the first to bring comparative order out of relative chaos. His classes and orders were clearly, if briefly, defined, the characteristics of his genera and species were well chosen and above all his method was simple to apply. To expect perfection in a pioneer effort of this kind is unrealistic, and critics of Linnaeus, who have often been heard, tend to forget this. Moreover, in his day evolution was not the accepted doctrine of biology; no more than faint whisperings of such a possibility were to be heard. Theorists still spoke of the Scala Naturae of the Greeks, and criticisms of Linnaeus were largely based on his divergence from ancient tradition.

More competent criticism came from Michel Adanson (1727–1806), a French botanist who was one of the most eloquent supporters of the idea that classifications could not be based on single, or even on a few characteristics, and that a more trustworthy method involved the observation and comparison of as many characteristics as was practicable. From the data thus aligned organisms might be grouped as indicated by the possession of a majority of shared characteristics.

Adanson was thus the first to perceive and to emphasize the fact that all the members of any one group need not show every one of its features. This was an extraordinary far-seeing hint of one of the main principles of modern taxonomy, and the device of including all or nearly all the observable details of the organisms in a group is sometimes known as Adansonism. It is in high favour with the practitioners of taxometrics. (See Chapter 8.)

Doubt as to the complete soundness of the Linnaean system was not immediately followed by attempts to improve it and the first serious revision appeared in Paris.

After the Revolution the Museum National d'Histoire Naturelle came into existence, a reconstruction of the former Cabinet du Roi, and the first election to the chair of invertebrate zoology was made in 1799. The professor appointed to take charge of 'Vers, insectes, et animaux microscopiques' was no less a person than Jean B. Lamarck (1744–1829). In the Museum Lamarck faced an immense task, but his genius was equal to the demands made upon it. He possessed an uncommon and enviable gift, an innate sense of classification, which he exploited to the full. His seven volumes of *Histoire Naturelle des Animaux sans Vertebres*, 1815–22, contains the results of his work.

In the first volume Lamarck discussed the process of evolution, and this led him to arrange animals from the Protozoa upwards, instead of in the reverse direction, from Man downwards, adopted by his predecessors. He took the controversial step of dividing the animal kingdom into three sections, distinguished by their mental capacities, and these were divided into classes, the hierarchy including the added step of family, suggested by Butschli in 1780.

Despite his personal interest in the possibility of evolutionary change, occurring as response to environmental demands, Lamarck's taxonomy was essentially static in nature. He accepted, and used, the idea of the type species, which persists in much practical taxonomy today owing to the help it gives in identification, but inevitably this put him in a somewhat paradoxical position. He subscribed to belief in divine creation and yet had reasons of his own to believe also in evolution. Nothing that he wrote seems to admit the existence of this contradiction, nor, in fact, does he himself seem to have resolved it.

A summary of Lamarck's classification is as follows:

Animaux Apathetiques
- Les Infusoires
- Les Polypes
- Les Radiares
- Les Tuniciers
- Les Vers

Animaux Sensibles
- Les Insectes
- Les Arachnides
- Les Crustacés
- Les Annelides
- Les Cirripèdes
- Les Conchifères
- Les Mollusques

Animaux Intelligens
- Les Poissons
- Les Reptiles
- Les Oiseaux
- Les Mammifères

Such an outline does not indicate the true value of Lamarck's contribution to the development of taxonomy, in its modern sense. He was the first zoologist to express definitely and to continue to support the theory that all the groups of animals in his classification had come into being by an evolutionary process. He believed that this process was a continuous one and that its apparent interruptions were no more than the result of our ignorance; and that in any grouping there should be detectable a rise from a lower, primitive, to a higher, more specialized form. The resemblance between Lamarck's deductions from contemporary knowledge and the opinion of present day systematists is too striking to require further emphasis.

Lamarck further showed a keen appreciation of the importance of precision in diagnosis. A high proportion of the generic and other names that he introduced are still in use at the present time.

From the point of view of systematics in general he was responsible for a outstanding innovation. Following his concept of evolution, he produced an illustrative diagram which displayed the groups of animals in the form of a branching tree. Here was the first entry of phylogeny into the schemes of taxonomy and the beginnings of its acceptance in the world of systematics.

All his life Lamarck was a contemporary and unfortunately a constant opponent of George Cuvier (1769–1832), holder of the chair of Anatomy at the central school of the Pantheon. He is usually regarded as the founder of comparative anatomy, on which a considerable part of taxonomy is based. He divided animals into four sections:

Animaux vertebres
Animaux mollusques
Animaux articules
Animaux rayonnés

a division that threw no light on phylogeny and had little to recommend it. Cuvier's greatest service to taxonomy was his insistence that extinct fossil forms should be found places in the tables of classification.

No doubt his work was embittered by his dislike of Lamarck, of whom he was lamentably jealous. The unequal labours of Lamarck and Cuvier mark the end of the pioneering period of animal taxonomy.

The following decades were marked by the emergence of three great ideas. The first was that usually known as Von Baer's Law, in which the author, professor at Königsberg, pointed to the fact that the younger an embryo was the more closely did it resemble other embryos of the same stage of development. The second was the explanation of this, offered by Ernst Haeckel of Jena. He suggested, in his Recapitulation Theory, that during its development an animal recapitulated the stages of its ancestral evolution, or, more briefly, that ontogeny repeats phylogeny. The third great theory was, of course, that of Natural Selection put forward by Darwin and Wallace in 1858 and expounded in *The Origin of Species* in 1859. Systematic zoology really dates from 1859.

Nevertheless, and with a sad irony, taxonomy fell into disrepute during the second half of the nineteenth century. Taxonomists did not find that the acceptance of evolution as an established fact necessitated any considerable change in their methods of classifying animals or in setting out their classification tables. The period, moreover, was marked by intensive collecting of animals and plants from all parts of the world, now more readily accessible than formerly; and enormous numbers of specimens flooded into the museums of Britain and Europe for examination, description and identification.

The systematists faced an almost impossible situation, bristling with insuperable difficulties. Schemes of classification were revised, abandoned and reconstructed, names were so multiplied that synonyms abounded, and only with single-minded devotion could enthusiam hope to survive. In consequence, taxonomists came to be considered as 'mere' museum men, who spent their days in a dead-end occupation, sorting and labelling the specimens set before them. No consideration was given to the fact that both sorting and labelling are in themselves important operations, that much work must be done before the label can be written, and that the preservation of specimens, often of historic value, demands constant attention. Without such care the ravages of mites and the constancy of evaporation will bring about the ruin of dry and spirit specimens respectively.

One reason for this was probably the surprisingly rapid growth of ecology and genetics and the contemporary progress in biochemistry. The zoologists of the early twentieth century flung themselves wholeheartedly into these new streams, leaving systematics to the minority.

Fortunately and inevitably, opinions changed. Soon after the First World War the field workers began to realize their dependance on the museum man. The consequence was almost spectacular and resulted in the appearance of *The New Systematics*, under the editorship of Sir Julian Huxley. The book became a landmark in the history of taxonomy.

The new systematics retains its adjective with the same conservatism in nomenclature that is found in the names of the New Forest and New College, Oxford. It was, first a restoration of the reputation of taxonomy, for the ecologists and explorers discovered that their results and conclusions lost most of their value if the animals on which they depended were wrongly named or incorrectly classified. Returning weary from their swamps and mountains, they had to appeal to the museum men to tell them what they had caught, a very salutory experience.

Simultaneously, the taxonomists realized that they could no longer neglect the knowledge that the geneticists, zoogeographers and physiologists were accumulating, the overall result being that a saner balance was established between the different branches of zoology.

In the writing of history the mistake of carrying the narrative too close to the present is one that can be and should be avoided. Nevertheless, and with recognition of the risk taken, this account cannot end without a reference to the state of taxonomy since about 1955, a state of such ebullience that it has been called the 'taxonomic explosion'.

The cause has been threefold. First, improved technique in the study of small objects, due first to the phase-contrast microscope and later to the electron microscope, has restored morphology to the position from which it had been temporarily deposed. The truth is now realized that whatever we may know about an animal's metabolism, genetics, behaviour or distribution, the only permanent record of material fact on which to base a classification is its structure.

In addition, the progress of biochemistry has been such that the molecular differences between species have become more fully realized, with a corresponding influence on our conception of the ranks of species, sub-species, races and varieties.

Finally, there can be no surprise that the computer has been called upon to make its peculiar contribution to the solving of our problems. It has made possible the wholly desirable introduction of quantitative analysis where previously visual comparison and personal opinion had held sway, and taxometrics promises to become one of the most rewarding techniques of the coming decades.

3

The Linnaean Hierarchy

The short account of the development of taxonomy given in the last chapter showed that the system in use today is founded on the work of Linnaeus. The essence of his plan in the *Systema Naturae* was an arrangement of groups of animals, or taxa as they are now called, in a series of increasingly inclusive range: similar species were associated in one genus, similar genera in an order, similar orders in a class. The principle was theoretically sound, as is evident from its survival and the virtual absence of any competing alternatives; in practice inevitable weaknessess were derived from various causes, such as the difficulty of deciding on the limits of each group and its status in the whole. But the system has the advantage that it is easy to put any known species of animal into its correct place, a fact which has often been compared to an address on an envelope.

Problems however, came later, and among the first were those caused by the increasing number of animals that became known as travel became easier and collecting more profitable. Linnaeus had divided the animal kingdom into the four groups named above, and these were sufficient for his purpose at the time at which he was writing. He was fortunate in that he knew, or had records of, only a small fraction of the animals of the world, for if he had had to deal with them all his genius, or that of any other systematist of the time, might well have been unequal to the task. For example, he knew seventy-eight species of Arachnida, and arranged them neatly in four genera—*Acarus*, *Phalangium*, *Aranea* and *Scorpio*—whereas between thirty and forty thousand confront his successors today.

Two consequences have followed. As a genus has grown in numbers it has become unwieldy and time-wasting in practice, so that it has been broken up into a number of smaller genera. This is helpful, but it has been accompanied by a corresponding increase in the size of the embracing family, which suffers the same treatment, and so on, up the list. This kind of operational revision continues to the present day.

With it there appears another limitation. As animals are more closely scrutinized, their differences become more and more obvious, until they become so impressive that the four steps of Linnaeus are too few to give them the emphasis that they seem to demand. The solution is to increase the number of steps, by the creation, interpolation and naming of more steps.

This was the first important elaboration of the Linnaean scheme. The term 'family' was introduced by Butschli in 1790 to bring related genera together, and the place it took, between genus and order, it has retained ever since. In the same way 'phylum' was used by Haeckel in 1886 to associate related classes, and in this instance there was a slight but important change in the point of view. The different phyla appeared either to be quite dissimilar, or to be so distantly related that common ancestors could not be detected.

The six taxa, phylum, class, order, family, genus, species, are known as obligate taxa, which means that every description of a new species of animal should in theory also indicate to which of the five superior steps it belongs. Sometimes this is too obvious to need actual mention, sometimes it is included in the general title of the describing paper.

More recently two other groupings have become widely recognized. They bear as names the familiar words 'cohort' and 'tribe': the former is interposed between class and order, the latter between family and genus. The useful word 'taxon', (plural taxa), meaning a group at any level between phylum and species, was formally adopted by the 14th Zoological Congress at Copenhagen in 1953, and has proved a most welcome addition to our vocabulary.

The literature of taxonomy contains a surprising number of other names that have been used by different writers at different times; they have been given no precise definitions and have never been generally adopted. These include:

Branch	Group	Phratry
Division	Legio or Legion	Proles
Form or Forma	Nation	Race
Gens	Phalanx	Section

Two of these, division and section, have occasionally been used with admissible if rather indefinite implications to denote a special relation between a small number of taxa, but in the present circumstances all the names in the list are best omitted from taxonomic work. A special case may be argued in favour of 'group' a convenient all-embracing word which may be regarded as the simple equivalent of taxon. It also appears sometimes when a genus contains a few species that are more closely allied to each other than to the other species in the genus, which, however, are not similarly grouped. This is not a practice to be encouraged.

The increase in the number of admitted taxa to eight does not complete the description of the hierarchy, nor has it proved to be sufficient to meet the demands of taxonomists. The traditional method by which the precision of the system can be improved is the addition of prefixes to the names of the taxa. Thus, a family might have its genera arranged in a number of sub-families, and might itself be a member of a super-family. There has never been any objection to this, and more recently the prefix infra has appeared below the sub-grouping.

This makes it possible to display the whole hierarchy, which, with four taxa to each of its obligate and its other steps, may now cover a range of thirty-two stages. (See page 22.)

There is probably no animal whose position requires for its complete and precise description the use of every one of these thirty-two taxa, but there are some that approach fairly close to such a state. If an example is given it is with the intention of showing that this seemingly elaborate range of steps has a real practical value and is not merely a piece of paper-taxonomy.

Kingdom
 Sub-kingdom
  Infra-kingdom
   Super-phylum
    Phylum
     Sub-phylum
      Super-class
       Class
        Sub-class
         Infra-class
          Super-cohort
           Cohort
            Sub-cohort
             Infra-cohort
              Super-order
               Order
                Sub-order
                 Infra-order
                  Super-family
                   Family
                    Sub-family
                     Infra-family
                      Super-tribe
                       Tribe
                        Sub-tribe
                         Infra-tribe
                          Super-genus
                           Genus
                            Sub-genus
                             Infra-genus
                              Super-species
                               Species
                                Sub-species
                                 Infra-species.

There is a strikingly beautiful orb-spinning spider to be found in Britain: it is a sub-species of the European *Araneus marmoreus*, and its full taxonomic achievement would read as follows:

Kingdom	Animalia
Sub-kingdom	Metazoa
Infra-kingdom	Triploblastica
Grade	Bilateria
Sub-grade	Coelomata
Infra-grade	Protostomia
Phylum	Arthropoda
Sub-phylum	Chelicerata
Class	Arachnida
Sub-class	Caulogastra
Super-order	Sternifera
Order	Arachnida
Sub-order	Araneomorphae
Super-family	Argiopiformia
Family	Argiopidae
Sub-family	Araneina
Genus	*Araneus*
Species	*marmoreus*
Sub-species	*pyramidatus*

Perusal of the above shows clearly how the hierarchic method of classification simplifies the placing of any animal in its correct position in the system, and simultaneously makes plain its relationship, as at present understood, to other animals. This process is one of identification, which must not be confused with classification.

When a number of unfamiliar animals are examined their resemblances and differences may suggest a system in which they can be helpfully arranged. The observations have come first, the system suggested has followed, a process that is describable as inductive or empirical classification. When an unfamiliar animal is examined and found to resemble others whose position in the hierarchy is already established, it is seen as to its genus and perhaps as to its species to identify itself with existing diagnoses. This is, of course, the common experience of every naturalist as he looks through the booty he has brought home; and, as has been said, its operation is made comparatively easy by the archetypal method, i.e. the use of a good key. This is a deductive process and is in the true sense of the term indentification.

The hierarchy that has been described in this chapter can lay claim both to the aristocracy of age and the reputation that follows many years of service. It provides a reasoned guide to the complexities of the animal kingdom, and has shown itself to be sufficiently adaptable to admit the countless new species and to absorb the changes in taxonomic opinion that have been presented to it during two centuries of zoological progress. Any criticisms must bear these facts in mind.

4

The Phylogenetic Aspect

One of the great advantages of the Linnaean or hierarchic classification is that it provides a ready means of deciding whether a seemingly strange animal is, in fact, 'new to science' or not. Should it be new, the system clearly indicates the place that it will hence-forward occupy. In performing this service the system manifests its relation to the *Scala Naturae* of Aristotle, and in its customary form it also makes a reasonable attempt to reflect the course of evolution. It is not a success for several reasons.

Fundamentally, it should be recalled that the purpose of a classification is one thing and that the indication of phylogeny is another, and that the purpose of classification is not to describe phylogeny. The purpose of building a house is not to demonstrate the properties of brick, stone and slate, but these properties are of basic importance and builders must take them into consideration.

The hierarchy tends to present too rigid a picture of a neatly arranged table, the components of which are a number of phyla that are more or less isolated from each other, and all of which can be easily divided into classes.

There may well be a hope that this does indeed represent truthfully the course of evolution, but the phylogeny of no group of animals is so perfectly known that all authorities agree on its probable course, and that of many, perhaps most, groups is very uncertain. Consequent gaps in our knowledge create a fictitious isolation of the phyla from one another, and this cannot give a likely picture of their past histories. Our conception of organic evolution is one of continuing and uninterrupted change, slow maybe, but having the result of relating all living organisms to ancestors from which, in the course of time, they have gradually diverged. Ideally, there should be no empty spaces or missing links.

There are so many well-defined phylogenetic groups that any form of classification that included them all would risk losing its value because of its very unwieldiness. Our system, in fact, is over-simplified and too optimistically organized. It seems to imply a single original ancestor for each taxon, which would then be described as a monophyletic group. Such an aspect of evolution is a naïve corollary of the Cell Theory of Schleiden and Schwann, for if it is true that every cell is the product of an earlier cell (*Omnis cellula e cellula*) then all the cells, and consequently all the bodies of all the animals alive today, must be traced back and back, through generation after generation, through age after age, to one unique, original and primal cell. This is not an unattractive picture, but it lacks probability and credibility. Doubt must be admitted as to its value, and with it the question whether some, at least, of our accepted groups are not more accurately to be described as polyphyletic. The implication is that they have each been derived not from one but from several ancestors, not necessarily contemporary.

The point may be elaborated by devoting a paragraph to the Protozoa, chosen because even an elementary introduction to zoology includes the genera *Amoeba* and *Paramecium*, and these are likely to be followed by mention of *Euglena*, *Vorticella*, *Monocystis*, *Polytoma*, *Plasmodium and Trypanosoma*, so that the Protozoa make a considerable impact on most students of zoology. But hear what authority has said about them. Professor H. Sandon wrote, 'The Protozoa show the untidy profusion of nature at its worst . . . and are correspondingly undefinable'; and from America comes a more emphatic expression of the opinion by Dr. Jahn that they 'are more or less one-celled organisms which are more or less animals'.

There are so many other phyla and classes in which the same 'untidy profusion' is evident, that an alternative, or perhaps a supplementary view of the animals of our world deserves attention. This is the well-supported opinion that the individuality of the phyla should be less emphasized than the existence of distinct, or at least distinguishable levels of construction or grades of organization.

The term 'grade' has thus entered the vocabulary of the taxonomists, and the idea associated with it is that of an assemblage of animal forms which share several common characteristics, but do not owe them to a common ancestor.

The Protozoa may again be called on, as they can throw light on the nature of a grade. Phylogenists who have sought for the origin of the Protozoa have made a number of well-supported suggestions. There are the possibilities that Protozoa have been derived from unicellular plants or Protophyta; or from bacteria by elaboration; or from cellular animals by degeneration. Any one of these may be true, all may be true of some of the Protozoa that we know today. Because the truth of one such hypothesis does not invalidate the possible truth of the others, the sole conclusion is that among the Protozoa the one common feature is that all its members have reached a stage of construction comparable to that of a single cell. In other words they represent the cellular grade of organization.

The traditional tree of evolution may need to be replaced or accompanied by another analogy. Perhaps the most picturesque is that of a bookcase with a number of shelves, and the image may be brought nearer to reality if on each shelf there is imagined to stand an array of objects, some branched, others single, some short and sad, others luxuriant and enthusiastic. And some are dead. In this picture the shelves represent the grades, and the objects thereon the phyla; and in the same way it must not be forgotten that there may well be classes that are themselves polyphyletic. A very obvious example is that of the Flagellata or Mastigophora, some members of which appear in zoology books as animals and in botany books as plants. The Sporozoa are also almost certainly polyphyletic.

A slightly elaborated version of this is shown in the accompanying figure.

The zoological bookcase is seen to have six shelves, only the highest of which is divided vertically. The two doors are shown opened. On the inside of the door on the left are the names of the established sub-kingdoms; on the right are the names of three

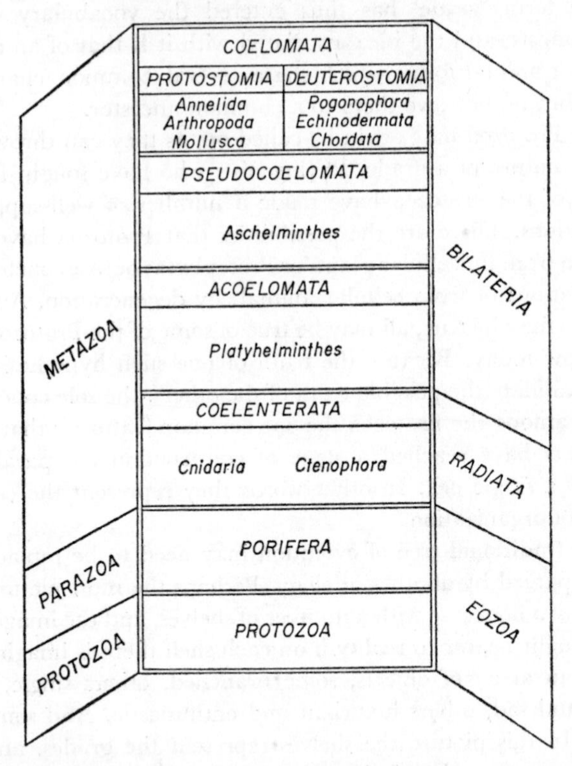

Fig. 1. A classification of the animal kingdom, designed in the form of a set of shelves, as an alternative to the traditional branching tree. The doors of the cupboard carry on the left the names of the sub-kingdoms, on the right the alternative names of the grades; the central shelves are named according to the sub-grades.

grades. The shelves are labelled with the names of the six most probable sub-grades, and the names of some of the contained phyla are also indicated. The whole is an attempt to diagrammatize one view of the relations between groups of animals, and it loses much of its value if it is not mentally compared with the traditional branching tree.

No more needs to be said about the bottom shelf, which holds the Protozoa. On the second shelf are the Parazoa, a sub-kingdom occupied only by the sponges. These are to be distinguished from higher animals by the fact that the cells are not dependent on one another, and they are to be described as a grade of cell-aggregates, clearly distinct from cellular animals.

The animals that occupy the third sub-kingdom, the Metazoa, possess bodies in which the cells are organized into tissues and organs and all co-operate to produce a unified whole. It is in this sub-kingdom that the idea of the grade is most acceptable.

At the beginning the cells are arranged in two layers, surrounding a single hollow. There is no true middle layer of cells (mesoderm) between the ectoderm and the endoderm: the cells form tissues, not organs, and hence their usual description Diploblastica. A feature common to these animals is that most of them are sessile and radially symmetrical. This symmetry contrasts so sharply with the bilateral symmetry to which we are all accustomed in more familiar animals that the name Radiata is given to the grade.

The highest form of animal body as yet evolved shows the cells in three layers and arranged in organs with special functions: this is therefore the organ level of construction, and since bilateral symmetry is established at the same time, the Bilateria are co-extensive with the Triploblastica.

The mesoderm is now present, and cells derived from it (mesenchyme) fill the space between body wall and alimentary canal. The important body cavity, the coelom, is as yet absent, so that the grade of organization bears the apt name of Acoelomata (or Acoela). The Platyhelminthes are the major representative phylum.

The formation and the functions of the coelom have no place in this book, and it must therefore suffice to distinguish between the 'false' and the 'true' coelom found in the animals not yet mentioned. In the former, mesodermal cells lie next to the ectoderm, but do not surround the alimentary canal. This is the condition in the sub-grade of Pseudocoelomata, in which some zoologists have recognized a number of phyla of obscure and inconspicuous creatures, showing an amazing mosaic of shared characteristics, while others have combined them all in a remarkable phylum, Aschelminthes, in which the familiar Nematoda are the most prominent class.

In the latter, mesoderm cells occur in both situations, somatic mesoderm below the ectoderm and splanchnic mesoderm outside the endoderm. There is a space between them, thus creating the final sub-grade of Coelomata. They may well be regarded as evidence of the fact that a body cavity was attended by such advantages that coeloms from two different sources have successfully established themselves.

Further analysis depends on the study of embryology, a science whose contribution to taxonomy has had a long and often a vital record. Von Baer's generalization about embryos and Haeckel's interpretation thereof in terms of recapitulation hold important places in the study of evolution, and the extension from embryonic to larval forms has been at least equally helpful to taxonomists. As examples in support of this there may be recalled the fact that the nauplius larva of barnacles showed that these animals were Crustacea and not Molluscs, that the cypris larva of the parasite *Sacculina* gave the same information, that the trochosphere larva is evidence of a commom ancestry for Annelida and Mollusca, and that the tornaria or dipleurula larva of *Balanoglossus* has guided us to place the ancestors of the Chordata among the Echinodermata.

Critical study and comparison of segmenting ova has revealed alternative fates for the blastopore. In some animals it marks the site of the future mouth, in others the site of the anus. To the former category, known as the Protostomia, belong the Annelida, Arthropoda and Mollusca; to the latter, known as the Deuterostomia, belong Pogonophora, Echinodermata and Chordata.

The significant aspect of the concept of grades is the manner in which it indicates different types of somatic organization. At each grade there appears the characteristic emergence of its chief phyla. Each phylum, probably polyphyletic, has developed its own potentialities, exploiting the opportunities offered by the environment, the different niches and habitats of which it invaded in the process we now describe as adaptive radiation.

Few questions in taxonomy are more often asked than 'What is the best classification?'. One answer, which deserves every consideration, must surely be based on the assumption that phylogeny is the most fitting theoretical basis on which taxonomy can be founded.

5

Taxonomic Practice

Many writers have said, and have said wisely, that there is no such thing as an impartial historian, since every written history has been coloured by the prejudices and pre-suppositions of the author. To this the historians may reply that there is no such thing as an impartial reader, since the impressions received are similarly coloured by the prejudices and pre-suppositions of the student. Herein, it is asserted, lies the attractiveness of history.

The parallel in taxonomy is not difficult to detect. There is no impartial systematist, nor are there any impartial users of taxonomic schemes; and only because the characteristics used in classifying must be judged and compared to determine their values or significance.

Classifications of animal groups are not made without long and painstaking study of many specimens of the animals concerned, and a sympathetic understanding of the conclusions so carefully reached may follow a short description of the actual work of a taxonomist on a single occasion. To do this, I quote a short paragraph, written a good many years ago, in appreciation of the unsolicited action of a former pupil. 'I am excited by finding in my laboratory this morning a small parcel, covered with foreign stamps and tied with unusual string. It is found to contain a large glass specimen tube, securely corked and protected by unintelligible newspaper, filled with red alcohol and containing the bodies of some dozens of exotic creatures. Accompanying this is a note: "Caught these spiders on Christmas Island and thought they might interest you. Good luck. C.G.–H." '

There is no doubt that the professional taxonomist would

tackle this collection in a routine manner that he has adopted for years past. It is also possible that some interest will be found, and some living spirit instilled into the activity of toxonomy, by a short description of the subsequent actions of one who, never recognized as a competent systematist, is sufficiently involved in the study of the animals themselves to wish to determine the exact nature of this unexpected present.

Certain preliminary conditions must be fulfilled before the collection can be examined with any chance of a worthwhile result. Even the amateur systematist will be expected to know and to be able to recognize the cosmopolitan species which the collection is sure to contain, as well as the narrower range of equally probable cosmotropical species. If these specimens can be quickly detected and at once removed, much time will be saved.

Then again, an almost absolute necessity is the possession of, or ready access to an adequate zoological library. To a general taxonomist, ready to attack specimens from any part of the animal kingdom (if any such person exists) this means no less than a library as extensive as that which can only be found in one of the great universities or museums.

This represents the chief advantage that the regular museum worker has over the enthusiastic but solitary amateur. The sources of information available to the latter are limited to such published books as deal with his subject on a reasonable scale, and, more valuable, his own collection of 'Separates', or the reprints of papers which scientific journals supply to their authors for distribution.

Probably the first step after the removal of the specimens whose identity is obvious is a preliminary sorting of the remainder into their appropriate or probable families. In many orders of invertebrates the family is readily detectable. After this the specimens will be taken individually and compared with the descriptions contained in the publications mentioned above.

One of three things will now happen to each specimen or group of similar specimens. It may be that the specimen will be found so closely to agree with the description that there is no doubt as to its identity. It can now be labelled and preserved.

Alternatively, the specimens may be found to be similar to a published description, but may possess some characteristics obvious to the investigator but not mentioned in print. To decide whether it is a different species or not the only certain course is to compare it with the actual specimen from which the first description was made. This specimen will have been chosen by its author from his original material and designated as 'the type' of the species and thereafter preserved with especial care in some recorded institution. Hence it is now necessary either to take the puzzling specimen there so as to make the comparison, or to try to borrow the type. Most museums, though valuing long series showing a range of variation, rightly refuse to allow type-specimens to leave their premises; but sometimes they can lend duplicates, technically called paratypes, which will serve the individual's purpose just as well. Or in some cases a curator may be willing to make the comparison and report on his findings if the specimen is forwarded to him.

Thirdly, the specimen may obviously be different from all those described in print or to be seen in collections, and in these circumstances it may assumed to be new to science, when a full description, accompanied by measurements and illustrations must be prepared for publication. Clearly, the process of deciding upon the identity of a new species is not a simple one, to be undertaken by the inexperienced.

An example of an unexpected difficulty may be mentioned. After the specimen has been compared with the dichotomic keys of the family concerned, it may be found to belong, shall we say, to the genus *Herpyllus* but not to resemble any of the species described under that name. But the genus *Herpyllus* has been created to accommodate a number of species from a growing genus *Drassus*, and it may be that this species has in fact been described as a member of the genus *Drassus*. Enquiry shows that the genus *Drassus* has a synonym *Drassodes*, and that *Herpyllus* itself has a synonym *Scotophaeus*. Therefore the newness of this species cannot be established until it has been compared with all the published descriptions of the genera *Drassus*, *Drassodes* and *Scotophaeus*. To which complication we may, for good measure, throw in the possibility that someone had perhaps desribed it as a species of a close genus *Gnaphosa*.

Let a return be made to the collection from Christmas Island. It was found to contain three small scorpions, 133 spiders and five false-scorpions. There was one earlier record of three collections from the island, published in 1888 and reporting the finding of one scorpion, nine spiders and three false-scorpions. Clearly the larger collection now in my hands might be expected to improve on this small number, but there was an additional point of more general interest.

The spider fauna of oceanic islands, such as Christmas Island which lies 200 miles south of Singapore, receives contributions from two sources. Some species accompany man and are carried on his ships and among his impedimenta, others arrive by air, floating on gossamer threads as part of the aerial plankton. Since Christmas Island is in the track of the prevailing winds from Australia it was desirable to notice whether its spiders were like those of Malaysia or those of Australia.

The scorpions were seen to belong to the species already described in 1888: the false-scorpions were forwarded to a recognized authority on the order, who acknowledged them with gratitude and no more was heard of them. From among the spiders seventy-one cosmpolitan and cosmotropical specimens were quickly removed and such literature as I had in my possession revealed the identities of a large proportion of the remainder. For the rest, it happened that by the fortunes of war I and my colleagues had been evacuated to Woodstock to continue our work in Blenheim Palace, a Hitler-given opportunity to seek help from the resources of Oxford. The literature and the collections in the Museums enabled me to solve all but one of the out-standing problems.

The remaining puzzle was represented by a couple of females which, to my great if undeserved satisfaction, showed at a glance an unusual feature: the lip between the maxillae was continuous with the sternum (figure 2). This was immediate evidence of their membership of a family named *Filistatidae*, in which alone this curious fusion exists. The family is a small one, but it happened that the Oxford material did not help me much. Because the family is predominantly an American one, further investigation

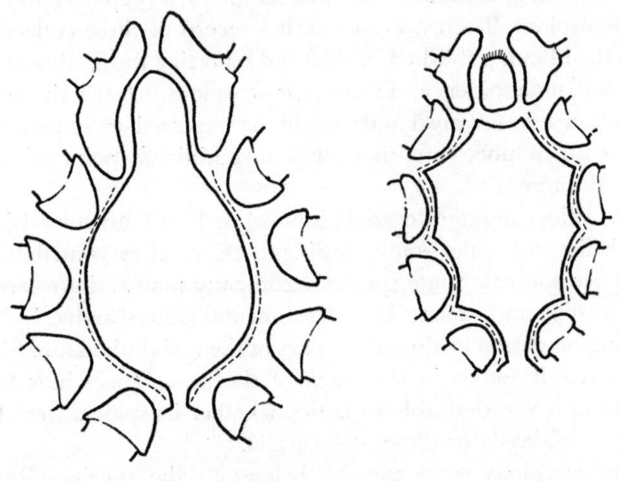

Fig. 2. United lip and sternum of Filistata (*left*) for comparison with separate lip and sternum of other spiders.

seemed to be worthwhile, and a visit to the library of the Zoological Society in Regents Park showed me all that I wanted. There was no doubt in my mind that these spiders really were a new species, and in due course they were so described. They were given the name of their captor. *Filistata gibsonhilli*, and with the rest of the collection were 'deposited' in the Hope Department of the Oxford Museums.

I have included this trifling narrative because it shows to some extent the amount of work involved in the examination of a collection of any animals from an unfamiliar region. Much depends, obviously, on the individual zoologist's experience of the group in question, as well as on the fact that some animals are easier to identify than others, and some taxonomists have a flair for spotting detail which many another might miss. On the whole, there should be no surprise that the report on any particular collection sometimes appears several years after it has been made.

6

The Species Problem

For many a year there was no 'species problem'. All men were content to recognize the fact that the world held animals of different kinds, moles and eagles, ferrets and crocodiles, tapeworms and kangaroos. The difference between the ordinary man and the zoologist was that the latter knew more about more kinds, that he called them species, and gave them Latin names. Meanwhile this satisfying position was maintained by faith in something called the 'fixity of species'. Fixity is an ugly word, and it means no more than constancy, but it gained popular acceptance. The idea was simple. An animal such as a ladybird might be known in zoology as *Coccinella septempunctata:* it was always and everywhere a small beetle, with two red wings-cases bearing seven black spots, and as such it was immediately recognizable. If the elytra were not red but yellow, if there were not seven spots but five, then the beetle was not *C. septempunctata.*

Such simplicity bordered on naïveté, but it was useful within its period. It recalls two other contemporary generalizations, the indestructibility of matter and the indivisibility of the atom. Both these beliefs have gone, and so has faith in the fixity of species.

When elaborated, this original idea implies that all species have been described and are now characterized by their structural details. This is often denoted by calling them morphological species, or, more shortly, morphospecies, and very generally they conform to the original idea of the nature of a species. This means that each species is quite clearly distinguishable from all other species, and that every specimen closely resembles the customary form or archetype, with no more than small individual or obviously abnormal variations. Further, no account is taken of evolutionary change, and all species are regarded as static or constant in their characteristics.

This emphasis on constancy may be condoned in view of the slowness of evolutionary changes compared with the length of human life, and in fact the concept of the fixity of species, that is of the morphospecies, has served the zoologist well for a considerable time, and continues to do so for a very large fraction of the animal kingdom.

Opinion began to change as travel and exploration became less difficult, and especially when the animals of any one group attracted intensive and critical study. In this respect birds take first place: their powers of flight, their brilliant colours, their songs, their behaviour, all these have combined to give them a fascination that can be challenged by few other groups. In consequence it was the ornithologists who took the lead in probing the fundamental question, 'What is a species?'.

The question arises when species previously recognized as distinct are found to be apparently connected by forms that are intermediate both in their geographical distribution and in their structural characteristics. Very fully documented examples are seen in the wren, *Troglodytes troglodytes*, in the British Isles; and the great tit, *Parus major*, all across the north temperate zone. Several other birds exhibit similar variations over wide areas, but birds are not the only animals that illustrate the phenomenon, for the existence of considerable variation within a species is to be found in every class in the animal kingdom.

There is, for example, a widely distributed primate called *Homo sapiens*, individuals of which show black skins and curly hair in Africa, brown skins and wavy hair in India, yellow skins and straight hair in China and white skins and wavy hair in Europe. This species is commonly accompanied by another, *Canis familiaris*, of which the Kennel Club recognizes over a hundred distinct types. One of the best examples is that of the long-tailed field mouse, *Apodemus sylvaticus*, which has a series of dark-coloured races in the Hebridean Islands: another is the bank vole, *Clethrionomys glareolus*, which has a curious little race on Skomer and another on the Channel Islands.

A first reaction to even so short a list as this would be to suggest that, variation notwithstanding, all these sub-species (or whatever they are to be called) proclaim the fact that they belong to one species because they interbreed and produce fertile offspring. No doubt this is true of all men of all colours, and of all dogs of all breeds, but who has shown that the whelks of the Baltic will breed with the whelks of the Mediterranean, or that the scorpions from Algiers will mate with the scorpions from Benghazi?

The contemporary view of this phenomenon is the announcement that the species is often a group, whose members are spread over a wide area in which considerable and constant variations occur. A species is therefore to be considered as made up of these different populations from different areas, and the population rather than the species is the unit of taxonomic study. All this is summarized in the word 'polytypic', applicable to all species that show geographically separate and morphologically different sub-species. Not only does it imply a contrast with the monotypic morphospecies that make so large a proportion of the animals that we know; it also proclaims an intense and widespread study of the species by many zoologists over many years. This is hailed as a great taxonomic advance.

The recognition of polytypic species forms a prelude to a faintly ludicrous position.

Zoologists, having accepted the superficial views of Plato, Aristotle and Linnaeus, had adopted the hypothesis of the fixity of species and now find themselves in the position of having to defend that hypothesis in the face of the contrary facts of nature. Such obstinate affection for an otiose hypothesis is not unknown in other branches of science. Both phlogiston and ether were a long time a-dying.

The first question that naturally arises evokes the preliminary qualification made famous by C. E. M. Joad, 'It depends what you mean by a species.' When this was put up to the systematists their answers boiled down to the words 'Really we don't quite know.' For a time the problem was avoided by an acceptance of Tate Regan's definition, 'A community whose distinctive morphological characters are in the opinion of a competent systematist sufficiently definite to entitle it to a specific name.' To this comforting sentiment the only serious objection is the lack of stated qualifications for competence among systematists.

A subsequent consequence was the introduction of a growing number of 'species', variously qualified. Some of these will be mentioned below. They represent the zoologists' struggle to retain the concept of species. It is mentioned in Chapter 10 that some critics of nomenclature remind us that zoology is not the study of animals' names: equally it were well to avoid the complaint that systematics is the classification of animals, not the creation of specific names. But criticism must admit that in all this the taxonomists are earnestly trying to replace a threat of chaos by an appearance of order, their basic function.

We proceed to a consideration of the nature of the animals described under the name of biospecies.

Within any genus of reasonable size there are to be found species that are quite distinct, some that are rather distinct, some that are rather similar and others that are very similar. In the last case the species may be indistinguishable by their structural characteristics alone, and have been given different names because they inhabit different types of environment or differ conspicuously in their habits. They are known as sibling species: an example is the pair of Drosophilas, *D. pseudo-obscura* and *D. persimilis*. One feature common to all these good species is that they do not normally interbreed. In other words their distinction is not a geographic one as in polytypic species, but is a genetic one. The species are genetically isolated. The two white butterflies familiar to so many naturalists, *Pieris brassicae* and *Pieris rapae*, are a good example.

This biological interpretation of the word 'species' has much to recommend it. It appeals to the essential nature of all living things, their capacity for reproduction, and it brings together beyond a doubt animals that may appear to be distinct. The difference between the sexes is in many invertebrates so great that many hundreds of 'species' of insects, arachnids and crustaceans have been founded on specimens of one sex only, and have been associated later, if at all, when observation of their behaviour has revealed the truth. Comparable differences may exist between larvae and adults: there is no more striking example of this than the Mexican axolotl, the paedogenetic larva of the amphibian *Ambystoma.*

In fact, many structural differences exist within a species, but there is no reason to believe that they necessarily constitute a barrier to reproduction: indeed they may even appear to encourage it, as any member of our own species can testify. Thus Mayr has thought to define 'species' as 'Groups of actually or potentially interbreeding populations, which are reproductively isolated from other such groups.'

These advances in ideas from the monotypic morphospecies of an earlier generation have enabled taxonomy to make more significant contributions to biology as a whole. The subject cannot be left without some reference to hybridization. It may well be true that good species do not normally interbreed, but it is a truism of science that there is often more to be learnt from apparent exceptions than from much repetition of the orthodox or conventional.

Hybrids are of at least two kinds. Artificial hybrids are often produced as a result of experiment. The loganberry is a good botanical example; so also is the progeny of a cabbage and a radish, which grows into a tree ten feet high. Among animals the outstanding example is the mule, a man-made marvel that is better adapted to certain human needs than either of its parents or any product of natural selection.

Other hybrids may occur in circumstances which do not involve the interference of man. Fishermen are quite familiar with the salmon-trout, a naturally produced hybrid between *Salmo salar*, the salmon, and *Salmo fario*, the trout. It is evident that external fertilization favours hybridization, for plenty of records exist of crosses between *Leuciscus rutilis*, the roach and

L. erythrophthalmus, the rudd; and between the rudd and *Abramis brama*, the bream. Whether these hybrids are themselves fertile seems to be uncertain.

However, all animals do not reproduce themselves by sexual methods, but increase their numbers by fragmentation, or by budding, or by parthenogenesis. Among the parthenogenetic animals there are some, like the Rotifers, in which no males have ever been found, and others, as among the Annelida, in which the individual is anatomically hermaphrodite but the male organs are functionless and produce no spermatozoa. Again, as in *Aphis* and *Phylloxera*, sexual and parthenogenetic generations are known; their appearance depends on the seasons and there can never be any doubt that the sexual and parthenogenetic castes belong to the same biospecies.

But among other animals, like the earthworm of the genus *Eiseniella*, the most abundant form, which is parthenogenetic, may be accompanied by several others, obviously closely allied to it in appearance, habits and habitat, and differing only in minor details, such as the precise segments covered by the clitellum. They may have been named as sub-species or as varieties, but a critical taxonomist seeks for more evidence for this. And the evidence is not easy to obtain: for if all the varieties interbred they would all belong to the same biospecies, and if they did not each would represent a different species. Because neither of these criteria can be used no precise answer can be given.

The suggestion may be put forward that all have come into being by mutation, thus postulating a single ancestor for the group. Other species have been produced by mutation, including particularly the sub-species of a polytypic species, but the essential difference of the parthenogenetic species remains. In sexual reproduction two individuals co-operate to produce the next generation; this is not true of parthenogenetic animals. In sexual reproduction the combinations of genes in the chromosomes of the two gametes involved makes mutation much more probable; in parthenogenetic species mutation must be much rarer, and the species must play a minor role in the process of evolution. All these ideas are summarized by describing such cases as agamospecies.

There have been many more species of animals in the past than are living in the world today, and the taxonomy of extinct forms raises problems of its own. Two chief features characterize the fossils that palaeontologists study: they are often fragmentary and in almost every case they are limited to the harder, more durable tissues. There is generally no evidence, or very little evidence, as to the nature of the softer, internal, organs. Secondly, they can often be arranged to show successive stages of a continuing evolutionary process; or, in other words, they introduce the dimension of time into taxonomy in a way that living animals cannot do.

The nature of the material makes it impossible to judge the remains on any other than morphological grounds, for obviously the criteria applied to biospecies and agamospecies cannot be used here: their morphological characteristics are a function of their geological ages, whereas those of living species are limited to those detectable at the present time. When, therefore, the taxonomist is confronted with a long series of fossil remains of such animals as Mollusca or Echinodermata he cannot be certain whether they are to be regarded as biospecies or as agamospecies. The one fact of which he is sure is that the individuals of the series are unlikely to have been contemporaries when they were alive. It is to such series of related forms that the name paleospecies is given.

In an attempt to summarize the foregoing account of 'species' and to make it more realistic, this chapter concludes with a fictitious survey of an imaginary animal belonging to a genus *Caborus* and inhabiting an area diagrammatized at figure 3.

1. At A and over much of the lake a species, named *Caborus deidatus*, is plentiful and was the first described form. At A2 the water is slightly warmer owing to protection from the westerly winds given by the wood, and slightly more acidic owing to the humus ashore. Here the form has been called *Caborus occidentalis*, as if it were a different species. At A3 the water is rather colder

owing to snow and water from the side of the hills. Here the form found has been called *Caborus orientalis*. In modern language *Caborus deidatus* is a polytypic species, a population showing three sub-species.

2. Some members of a very similar kind have taken to living on land. At B1 they are found living a cryptozoic life under rotting leaves and in the humus. Specimens at B2 live under

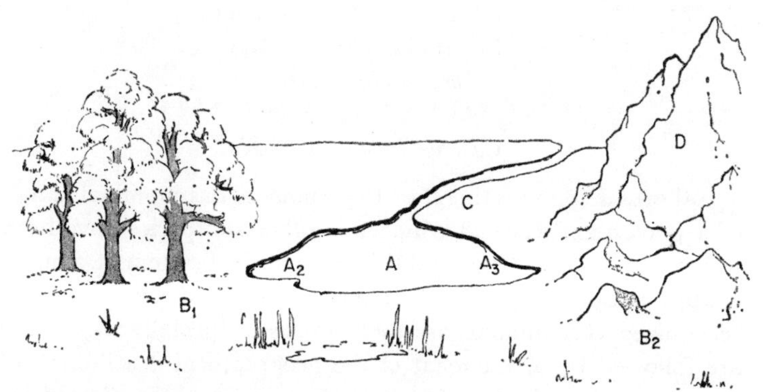

Fig. 3. Various habitats of Caborus deidatus.

stones and rock fragments: both are thus protected from exposure and desiccation. The B1 specimens are rather smaller and darker in colour, otherwise the two populations are very similar. If they met, which has never occurred, they would almost certainly breed together. They form a biospecies, and have been called *Caborus terrestris*.

3. Members of a third form have migrated to the river and have survived in the running and more highly oxygenated water at C. They probably owe their success to a mutation, which has also had the effect of making them parthenogenetic. Named *Caborus fluviatilis*, their relation to the lake-dwelling forms is indeterminable, and they form an agamospecies.

4. No specimens from B2 have been able to ascend the hills (D), but the rocks flake easily into slabs in which by good fortune there are many fossils, chiefly of carapaces and limbs of an

ancestral species, very closely allied to those alive today. Very probably they could have mated with these if they had met, and they may be described as a paleospecies, *C. monticola.*

What is to be made of the foregoing? One answer to this question is that by the generous tendency of the present we may look upon the whole group as a superspecies composed of the following:

Caborus deidatus deidatus
Caborus deidatus occidentatus
Caborus deidatus orientatus
Caborus deidatus terrestris
Caborus deidatus fluviatilis
Caborus deidatus monticola

Caborus deidatus is therefore the binomen (or binominal name) of a polytypic species of which five other subspecies have been distinguished. The long established binominal system of nomenclature is not adequate when populations and subspecies are recognized. Trinominals are more precise, especially when they are followed by a statement of the geographical distribution of the variety, but they tend to be clumsy and to be neglected for this reason by some zoologists. Perhaps they suggest a reversion to the antique system in which an animal was named not by words but by a long phrase.

The basic idea of the species with its binomen is so well established by tradition and so efficient for most of the known animals that the modifications discussed above must spread but slowly into the bulk of the animal kingdom. But in adopting it one should remember that it is an attempt to give expression to the facts as they appear to the investigator and to demonstrate the relationships between the animals concerned.

7

Doubts and Discontents

After the discussion in the last chapter as to the nature of a species the questions that logically follow must be concerned with the nature of a genus.

The problems here are fundamentally different. The distinction between species based on adaptations, behaviour and fertility are the natural consequences of selection, while a genus is a grouping of similar species which are assumed to be related because of their similarity. It is, in fact, no more than an invention of taxonomists, a device that is useful to them in their work. Doubt first arises when the number of species that a genus may usefully contain comes to be discussed. The question, 'How big is a genus?' is one of the same calibre as the more familiar, 'How long is a piece of string?' but answers can be suggested from both a practical and a theoretical point of view. Many a teacher has found that the ideal size for a Sixth Form is twelve, and that fifteen is bearable, while in the Middle School twenty-four is ideal and thirty is tolerable. It seems reasonable to suggest that in some families a limit of twelve or fifteen may be set to a genus, in other families these figures may be replaced by twenty-four and thirty. When a genus threatens to exceed this number revision by a specialist is justified. This is the practical answer.

The theoretical answer concerns the grading of ranks in the heirarchy, or the determination of the place to be assigned to any particular grouping of animals. The pertinent questions are therefore 'What distinguishes a family from an order: where is the line of demarcation between an order and a class?'

The uncertainty arises from the essentially artificial nature of the hierarchic system, for herein lies one of the outstanding features of taxonomy. Animals have evolved, but their evolution has been without purpose and has not been directed towards the creation of a convenient system of classification. It has been said before, and it can scarcely be repeated too often, that just as nature made the integers and man made mathematics, so too nature made animals and man made zoology, and in particular made taxonomy.

There is, therefore, no rigid structure, no rule that proclaims 'Here this genus stops' or 'Now this family has become so diverse that it must become an order'. All decisions reached are expressions of the personal opinion of the taxonomists who are working on the group. Since, as we are daily made aware, opinions differ, there are sure to be disagreements and apparent inconsistencies all through the classification table; but it is this opportunity for the exercise of personal choice that justifies the claim of taxonomy to be considered as an art. One of the tasks of an artist is to convey his impressions of nature to others, and this is exactly what a taxonomist must try to achieve in the construction of his tables of hierarchic taxa.

There are, no doubt, many orthodox systematists who would take a different view of the nature of a genus, and give it as real an identity as they give to species. They would base their belief on two arguments. First, they would say that species at one end of the hierarchy and phyla at the other are universally and reasonably accepted as expressions of the course of organic evolution. Therefore it is illogical to withold from the intermediate ranks the status that is accorded to either extreme.

Secondly they would point to the fact that prolonged study has shown that species do fall into natural groups, which may surely be called genera. The implication of the word 'natural' is uncertain: it may be no more than the recognition of gaps in a series, which are very possibly present because the intermediate species have become extinct. They might add that the computers used by the exponents of taxometrics seem to point to the apparent reality of the genus. (See Chapter 8.)

Differences of opinion are as active in taxonomy as in art or politics, and the consequences are sometimes surprising.

For example, in the family Liphistiidae there are two genera, *Liphistius* and *Anadiastothele*. At the hands of one authority the latter was abandoned and its species merged with *Liphistius*, while another authority not only retained the genus but gave it the distinction of putting it into a separate sub-family by itself.

The inclusive nature of families gives them a greater stability than genera, and a family is less likely to suffer extinctions than a genus. This continues with increasing validity to higher ranks; orders, classes and phyla are successively less liable to modification in diagnosis or composition. They are sometimes robbed of a component when one of their taxa is extracted and promoted to higher level. Examples of this will be found below.

Changes of this kind are not the result of passive rumination, but have been suggested by intense scrutiny of the animals themselves. In the nature of things they are almost always dead, preserved specimens, and the taxonomist must limit himself to their external and immediately obvious characteristics. To accept this limitation is, as a rule, the only practicable course in the circumstances in which the specialists in museums have to work; yet they, like other zoologists, realize that external features alone cannot tell them the whole story of an animal's evolution.

As against this, there are disadvantages in attempting to look for further information among the internal organs. Dissection takes a long time and demands a considerable degree of skill; section cutting requires a more elaborate technique, and examination of the sections takes almost as much time as does the dissection of a larger animal.

Again, the animal under examination may be so rare that few specimens are available, and the taxonomist may naturally hesitate before cutting one of them into small pieces. Even if he does so, later workers, who may have to compare their specimens with his descriptions or drawings, may not have the wish or the skill to carry out the necessary dissection. For the present, therefore, and probably for many years to come, our classification of animals is likely to depend on the external features of their dead bodies.

All this is only another way of saying that animals are compared and contrasted from a qualitative rather than from a metrical, quantitative, or numerical point of view, and one of the most conspicuous results of this is that systems of classification have always been bedevilled by the opposing methods of the 'splitters' and the 'lumpers'.

These inelegant but universal descriptions have the advantage of being self-explanatory, and of dividing zoologists into those with an analytical mind, who tends to split taxa into smaller units, and those with inclusive minds, who tend to remove barriers and distinctions wherever intermediate forms make this possible. There are arguments in favour of both opinions.

One of the justifications for the lumpers dates back to the days when the very truth of the doctrine of evolution was being discussed. When, for example, the suggestion was made that the classes of fishes and amphibians should form one group with the name Ichthyopsidea, or that the reptiles and birds should similarly form one group called the Sauropsidea, these names carried a suggestion with them. They are not widely used today, but anyone who adopted them in the past was implicitly stating his belief in an evolutionary relation between the classes.

Although the tendency to lump species together is increasing, the reverse is true of orders and classes. Without difficulty, without searching the specialists' journals for their treatment of the rarer animals, following instances may be noted:

(i) The phylum Coelenterata used to include the class of free-swimming comb-jellies or Ctenophora. Recently these have been raised to the rank of a phylum and the residual sessile forms have been named as the phylum Cnidaria.

(ii) The class Myriapoda contained two sub-classes, the Chilopoda or centipedes and Diplopoda or millipedes. The differences between these two groups are now held to be great enough to justify putting them into two separate classes. These retain their former names.

(iii) The fishes were for many years regarded as constituting a single class with the classical name of Pisces. Their relative uniformity was most probably a consequence of convergent evolution, due to the fact that they all lived in the same sort of environment, the water. Investigation, supported by increased knowledge of fossil forms, has suggested that the fishes are a typical polyphyletic group and at least two classes, Chondrichthyes and Osteichthyes, are now listed.

(iv) The insect order Orthoptera was always one of the largest in the class Insecta, and its members today occupy the four orders of Dictyoptera for the cockroaches, Dermaptera for the earwigs, Chelentoptera for the stick-insects and Orthoptera, sensu stricto, for the locusts, grasshoppers and crickets.

(v) The order Rodentia has shed the rabbits and hares, which now constitute an order Lagomorpha.

These examples show that while among the higher taxa the dividing tendency has much to recommend it, the splitting of the lower ones is not always so acceptable. When the lamellibranch genus *Inoceremus* was converted into two families, with twenty-four sub-families, sixty-three genera and twenty-seven sub-genera, it was difficult to believe that the cause of taxonomy was being advanced.

In general terms, the dividing of any taxon has one of two causes: either it is due to the increased number of species or genera which seem to find places in it, or it is a result of more and more detailed scrutiny of the animals themselves. Inevitably the discovery of more and more animals calls for more meticulous examination of both the newer and the older specimens, and the fundamental fact is that to the investigator the observable differences make a more powerful impact than the resemblances. This basic feature of the mind was mentioned with an experimental example in Chapter 1; it is largely responsible for the way in which the taxonomy of animals has developed over the years.

Justification has already been sought for the limiting of the examination of animals to their external, easily visible characteristics, but support for the alternative view is so clear as to require consideration. Such limitation is logically indefensible, for there is no reason to believe that the internal organ systems vary in correlation with the outer surface, nor that these systems themselves have not undergone their own evolutionary changes; nor indeed that a classification based on internal structure would necessarily resemble the accustomed scheme based on externals.

In consequence, very considerable interest attaches to such attempts as have been made to build a classification on a combination of internal and external characteristics. One of the most striking of these was the *Inquiry into the Natural Classification of Spiders*, published in 1933 by Petrunkevitch of Yale. He had cut sections of representative species of almost every family of spiders and had found that they could be logically arranged in a system based in the first instance on the number of ostia between the heart and the pericardium. Combining this clue with other details of structure, both internal and external, he offered a revised classification of the order, which seemed truly to deserve the description of 'natural'. Its publication received well-merited and world-wide acclamation; it was described as a great step forward in the principles of taxonomy, its methods were recommended to all zoologists, and were actually put into practice* by several specialists in other groups. But what promised to introduce a taxonomic revolution failed in practice to do so. And the reason was obvious. The new 'natural system' was admirable in theory, but the practical araneists who were called upon to examine collections from any part of the world continued, excusably, to use the well-tried traditional methods. They may not have been phylogenetically ideal, but at least they could be used and understood.

Phylogeny, though based on comparative anatomy, requires palaeontology both to corroborate its ideas and to suggest new ones; from which it appears to be quite inevitable that extinct taxa, at all levels from genus to phylum, ought to be given places

* Any who think of following this example should perhaps be warned that Petrunkevitch had made 9500 slides in coming to his conclusions.

in our classifications. The first result of this is, of course, an increase in the number of groups with which the taxonomist is called upon to deal, and often the increase is considerable. The truth is that most of the species of animals which have lived on the earth are now extinct: the neozoologist may be heard to complain of his million species or more, but the palaeozoologist may retort that he is fortunate to have to think of so few. For example, Simpson tells us that 54 per cent of the families and 67 per cent of the genera of Mammalia are known only as fossils; and by the same token the class Arachnida grows from eleven living orders to a total of sixteen, an increase of 45 per cent.

When a palaeontologist turns his attention to uniting the classifications of both living and extinct forms, the result is sometimes startling. Scorpions provide a good example. The simple grouping, to be found in any general zoology text, divides the order into six well-defined families, and there is no doubt that for most ordinary purposes this is quite sufficient to enable a student to gain a clear grasp of the characters of the scorpions living in the world today. But in a treatise on invertebrate palaeontology the same students will find a classification divided into two sub-orders, nine super-families and twelve families.

If increases like these are found in other classes of the hierarchy, a comprehensive classification of the animal kingdom would fill a very considerable volume; and it would, moreover, become a work of very occasional reference, consulted by a number of zoologists who could be counted on the fingers of a mutilated hand. But taxonomy is a vital study, demanding more attention than follows a neglect consequent upon its own elaborations.

This introduces a problem of a different nature, and one which the pure scientist may consider to be rather irrelevant to his work, concerning the manner of its presentation on the printed page. It may be said that this is solely the business of the type-setter, but the taxonomist's point of view should not be overlooked. Alternatives exist and are used indifferently, but seldom does the zoologist recommend a standard to which others may conform or merely approach. Once again personal opinion or, as it is also called, private enterprise finds a place in taxonomy where it is subject to no authority.

Consider the following example.

(1)

Insecta
Apterygota
Pterygota
Exopterygota
Endopterygota

(2)

I Insecta
A Apterygota
B Pterygota
a Exopterygota
b Endopterygota

(3)

Insecta (= Hexapoda)
Apterygota (= Ametabola)
Exopterygota (= Hemimetabola)
Endopterygota (= Holometabola)

(4)

Class Insecta
Sub-class Apterygota
Sub-class Pterygota
Infra-class Exopterygota
Infra-class Endopterygota

(5)

Class Insecta Linnaeus, 1758
Sub-class Apterygota Brauer, 1885
Sub-class Pterygota Brauer, 1885
Infra-class Exopterygota Sharpe, 1899
Super-order Palaeoptera Martynov, 1920
Super-order Neoptera Martynov, 1925
Supra-order Orthoptoidea Martynov,
Supra-order Hemipteroidea Martynov,
Infra-class Endopterygota Sharp, 1899

The objection to the above (that it is a question for the printer) is true, but the uncertainty is a sympton of the general malaise that spreads through taxonomy and may appear almost anywhere at any time.

Even a casual examination of the literature of systematics during the past half-century reveals a surprising number of suggestions, some of which are sufficiently surprising in themselves. There have been suggestions for a completely new 'non-Linnaean' system, in which organisms are somehow plotted between Cartesian co-ordinates, and appear as points, clustered or scattered. There is, too, much despair at the difficulty of including phylogenetic relationships in tables of classification, and this has led some taxonomists to express a wish that phylogeny be altogether omitted from consideration.

There is always dissatisfaction with the binominal nomenclature, which some writers have wished to replace by a uninominal system, in which every animal had a name of only one word. One trembles to think of some of the 'words' that might be invented. Others believe that advantages would follow if every species were known by a number; and one may imagine the proposer of this idea telling his colleague that he had that morning heard the song of a 279·63, or that he had been painfully bitten by a 00968·51. Small wonder that in 1907 von Wettstein suggested that recognition should be given to two distinct systems of classification, one for truly scientific research and another for amateurs, nature lovers and other laymen.

This aspect of taxonomy is in sharp contrast with the stability of other scientific systems such as the Periodic Table or the Laws of Thermodynamics. It is, no doubt, a consequence of the differences between living, variable organisms and lifeless, inert matter; and the fundamental cause of all dissatisfaction was clearly described by Kiriakoff in 1961. He wrote 'I believe that the only truly scientific attitude is to confess that the importance we attach to systematic arrangements, ordering of groups, hierarchical value of taxa, etc. is not justified.' The system to be used is practical, and may contain data from phylogeny, but most of it is purely conventional.

This leaves the matter in the position of admitting that the best system to use may be a purely practical one, since so much of the taxonomy must always be merely conventional. But this is a rather defeatist attitude, to which no scientifically-minded biologist could be expected readily to subscribe.

8

Numerical Taxonomy

Much of the discontent with the present state of taxonomy can be traced to the uncertainties that inevitably follow a qualitative and subjective study of any kind. The purely scientific mind is impelled to seek for quantitative methods which, it will be believed, will remove all doubts about the weighting of any one characteristic, will determine the limits of any taxon and indicate its rank. An ideal system of classification would be based on resemblances between organisms, would make use of as many characteristics as seemed desirable, and would use numbers to indicate them rather than descriptions. Such a system, being quantitative, would be scientific in the sense that any two systematists using it to construct a classification of any group of organisms must inevitably produce identical results. This would be in sharp contrast to the conditions that obtain at present, in which two systematists would quite certainly arrive at different conclusions.

The present is the age of the computer, and the rise of numerical taxonomy, or taxometrics, since its first appearance in 1957 has been entirely due to the powers of these instruments. At the same time it has evoked a good deal of valuable discussion which has been fortunately free from any trace of ill-tempered controversy. Manifestly, numerical taxonomy has still a long way to go and will have to appeal to a larger number of zoologists than at present. In this chapter an attempt is made to explain some of its basic principals in simple terms.

The first necessity is to decide on the characteristics that shall be used. These are often referred to as 'unit characters', a phrase which implies that they should be of an all-or-none nature, something that is either present or absent. Alternatively, they should be able to be measured or counted or evaluated in terms of an arbitrary scale. As an example of the last-named the stoutness of a leg may be taken. An invertebrate's leg may be described as very slender, as thin, as normal, as stout, or as very stout, but it is more satisfactory to measure the tibial index, in which the diameter of the tibia is expressed as a percentage of the combined length of the tibia and patella.

Let this now be applied to two animals, A and B, of which it can be observed that:

A has four eyes and six legs
B has six legs and eight eyes

These four facts can be plotted between two co-ordinates, as in figure 4, and the distance between them can be easily calculated.

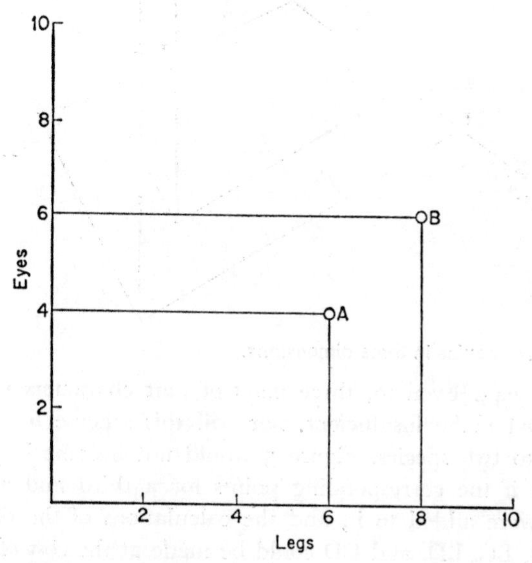

Fig. 4 Taxometrics in two dimensions.

It is $\sqrt{8}$, or 2·83: this is a measure of their relationship and by itself is almost meaningless.

But suppose, further, that A has two pairs of mouth-parts and B has three pairs. This introduces a third dimension, as in figure 5. It is still easy to calculate the distance between A and B: it is,

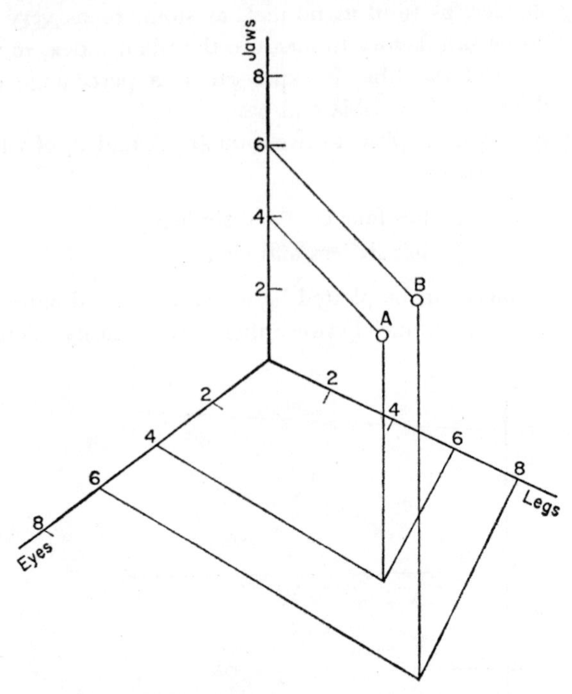

Fig. 5. Taxometrics in three dimensions.

in fact, 3·43. Even so, three pairs of unit characters must be considered to be insufficient, nor will this scheme have to be limited to two species. Figure 5 would not become impossibly complex if the corresponding points for a third and a fourth species were added to it, and the calculations of the distances AC, AD, BC, BD, and CD could be made at the cost of a little time.

However, when the number of pairs of unit characters is increased beyond three, by including, for example, the numbers of ganglia in the ventral nerve cord and the number of subdivisions of the tarsus and the number of ostia in the percardium, the six pairs of characters could only be plotted in multidimensional space or hyperspace. The distances between all the points now present can no doubt be calculated by the ability of the mathematical mind, but the time taken would be prohibitive. Here the computer is used and for this reason numerical taxonomy did not progress beyond the theoretical state until computers were available to provide us with the figures required. How are these figures to be used?

Again an imaginary case will be offered to illustrate the process. Consider a group of twelve organisms, A to L, which have been subjected to this treatment with respect of a sufficient number of characteristics. The results can be tabulated in a form known as a similarity matrix, shown in figure 6, where the actual

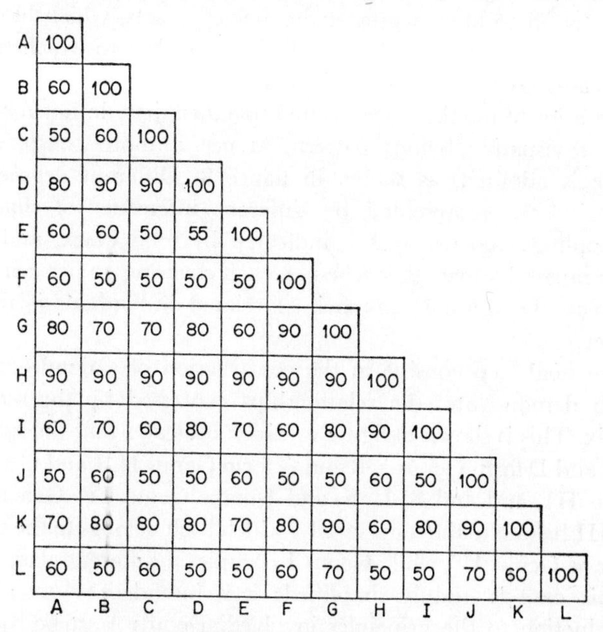

	A	B	C	D	E	F	G	H	I	J	K	L
A	100											
B	60	100										
C	50	60	100									
D	80	90	90	100								
E	60	60	50	55	100							
F	60	50	50	50	50	100						
G	80	70	70	80	60	90	100					
H	90	90	90	90	90	90	90	100				
I	60	70	60	80	70	60	80	90	100			
J	50	60	50	50	60	50	50	60	50	100		
K	70	80	80	80	70	80	90	60	80	90	100	
L	60	50	60	50	50	60	70	50	50	70	60	100

Fig. 6. Similarity matrix for twelve species: first stage.

numbers are replaced by percentage values. Thus the 100 per

cent occurring along the diagonal side indicates that A is (obviously) exactly like A, B like B, and so on, whereas in the third line the resemblance between A and C is 50 per cent, between B and C is 60 per cent etc.

In the next step the matrix is rearranged to bring together the species that most closely resemble each other. There are several ways in which this can be done: in this example the elementary, primitive, but easily understood course has been taken of calculating the average values along each horizontal line. For less imaginary cases the expert taxonomist, who, it will by now be realized, must also be a mathematician with a knowledge of statistics, has to choose the method that best suits his needs.

When the regrouping has been performed the organisms A to L are found to have arranged themselves in sets, which in conventional taxonomy would be called taxons, but to which taxometricians give the name of phenons. This is shown in figure 7. When a lot of numbers are printed together they do not tend to form a visually obvious pattern. Hence a more conspicuous device is adopted, as shown in figure 8. Different degrees of similarity are represented by different intensities of shading or stippling: 100 per cent is indicated by solid black, and the lower ranges by five other schemes, each covering 10 per cent. In this way the mutual approach of related individuals is made clearer.

The final step consists in the construction of a dendrogram, which demonstrates the relationships expressed by the second matrix. This is shown in figure 9, which indicates that the species A, H and D form Genus I, B and G form Genus II, I and C are in Genus III, and and E, L, F, and J make Genus IV. Genera II and III belong to the same family, which itself belongs to the same order as Genus IV, while Genus I occupies a different class.

This case is unduly simplified: it is intended to act as an introduction to the principles involved. Details must be sought in larger volumes than this.

At this stage it might be well to point out that the exponents

	A	H	D	B	G	K	I	C	E	L	F	J
A	100											
H	90	100										
D	80	90	100									
B	60	90	90	100								
G	80	90	80	70	100							
K	70	60	50	80	70	100						
I	60	90	80	70	60	60	100					
C	50	90	90	60	70	80	60	100				
E	60	90	55	60	60	70	70	50	100			
L	60	50	50	50	70	60	50	60	60	100		
F	60	90	50	50	90	60	80	50	50	60	100	
J	50	60	50	60	50	90	50	50	60	70	60	100

Fig. 7. Similarity matrix for twelve species: second stage.

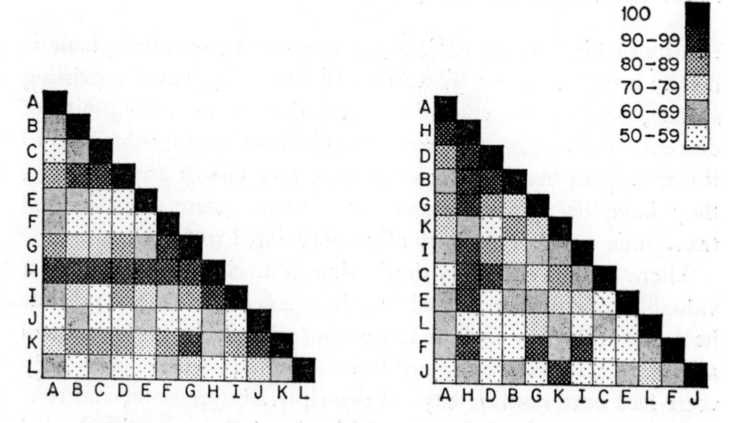

Fig. 8. Pictorial, shaded, form of matrices in Figs. 6 and 7.

of numerical taxonomy are accustomed to use a number of terms that are unfamiliar to other taxonomists. For example, classifications based on the presence or absence of a single or only a few characteristics are described as monothetic, in contrast to the preferred classifications which are polythetic. Again, the

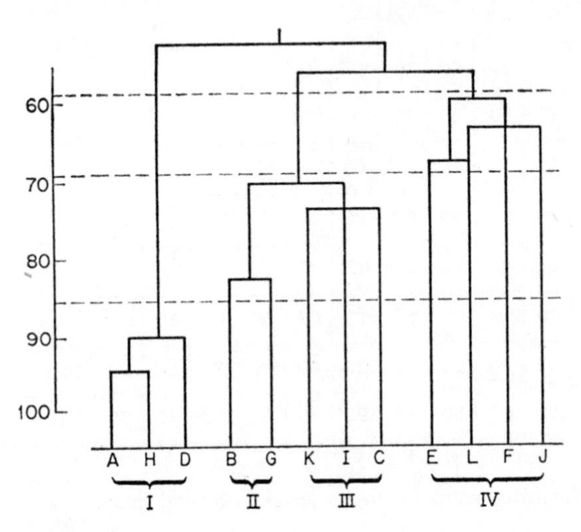

Fig. 9. Dendrogram of twelve species.

relations which taxometricians try to evaluate are called phenetic if they are determined by an overall similarity between existing species, or they are cladistic if they depend on community of descent. In the same way, since the objects that are submitted to this treatment may be species or genera or almost anything else, they have been given a general covering term, 'operational taxonomic units'. This is usually abbreviated to OTU.

There can never be any doubt that numerical taxonomy has a valuable contribution to make to the science of biology, and that its value will grow as it is more appreciated. Its basic features are that it introduces quantitative methods where previously there had been nothing beyond descriptive impressions, and that by so doing it eliminates the subjective influence of personal prejudice that has done so much to retard the progress of classification. In particular, it is claimed that its phenetic principles avoid the errors that follow over emphasis on, or over-weighting of, one characteristic in particular. Further, it helps to decide the limits of a genus, a family or other taxon.

The advantages thus claimed may be summarized by saying that taxometrics is a device that changes conventional taxonomy from an art to a science. If this is true, and there are those who regretfully admit that it may be, no opposition to its establishment is likely to last for long. For opposition there has been, and discussion on various points will continue, though it will be concerned with procedure rather than with the fundamental principle. For example, in taxometrics the number of characteristics involved is often quite considerable, the first belief being that the more characteristics included the more trustworthy the result must be. But unlimited increase is not to be countenanced, and the optimum number demands discussion.

Numerical taxonomy must not be misunderstood. It is not a method of identification. One cannot take a strange animal, count its parts and measure its bits and pieces, feed the resultant figures into a computer and receive an answer such as '*Anelasmocephalus cambridgei* West! But whatever its limitations there remains the fact that taxometrics represents a new approach to problems of classification. Perhaps it foretells a revolution, or, in modern parlance, a breakthrough; perhaps it does not. But for the taxonomists it represents a new technique, and new techniques have brought about unforeseen advances in so many other branches of science that they teach us a lesson from which it may not be unduly optimistic to hope for much.

9

Chemical Taxonomy

There have been times in the history of biology when controversy has arisen between two opposing schools of thought, known as the mechanists and the vitalists. Very briefly, the former held the opinion that all events in the living organism should be explicable in terms of chemistry and physics, and that only our ignorance prevented us from turning biology into a branch of chemistry; while the latter denied this and asserted their confident belief that in life there is 'something more' than chemical reactions and physical change.

Arguments of this kind are seldom heard today, probably because the vitalists can find no new arguments with which to support their belief, but the rapid progress of biochemistry is felt in widening spheres and cannot be neglected by taxonomists.

It was during the Second World War that the efficiency as an insecticide of dichloro-diphenyl-trichlorethane, or DDT, was first exploited on a large scale. It seemed to be fatal to every flea, every mosquito, every louse against which it was directed. Certainly it killed a very high proportion of any of these pests; equally certainly a few survived. Their survival was 'explained' by saying that they had a natural immunity, and when they produced offspring that had inherited this immunity the economic entomologists were faced by a 'resistant strain'.

Clearly enough, a different reaction to the same chemical stimulus can only indicate a different chemical reactor; in other words in the tissues of a number of individuals of the same species the chemical composition was not uniformly constant. Two questions arise: how to detect the natures of the differing compounds, and what contribution can such knowledge make to the science of taxonomy. Fundamentally it must seem that considerable contributions may be expected, for the metabolism of an organism is a complex of chemical changes (whether or not there is 'something more'), and all the morphology, behaviour and ecology of an organism must depend on its metabolism. Biochemical taxonomy was first mentioned, though not under that name, by A. P. de Candolle at the beginning of the nineteenth century in a short reference to some interesting chemical relationships between different species in the same genus of plants. He drew attention, for example, to the presence of terpenes in all species of *Pinus*, to the fact that all species of *Cinchona* were useful in cases of fever, while all members of the Convolvulaceae had laxative properties.

Biochemical taxonomy was, in fact, at first a botanical monopoly, just as were the first movements towards an official code of nomenclature and, later, to the establishment of the science of ecology. Zoologists have followed the examples thus set, and are sometimes forgetful of the debt they owe to the guidance of their fellows.

At the time of de Candolle's hints biologists were not in a position to follow up the significance of the isolated facts. Evolution was scarcely an accepted theory, and, more important, the necessary laboratory techniques had not been conceived. But when, a hundred and more years later, these obstacles were removed by far the greatest progress came from the labours of the botanists. In nearly all the published work emphasis is almost wholly on plant tissues, with references to animals in a very small minority. It is therefore important that zoologists should be introduced to the possibilities that exist for themselves.

Maintaining the principle of making a start with very simple cases, one may recall one of the first facts impressed on the student of biology, the occurrence in plants of a cellulose cell-wall. No such wall surrounds the cells of an animal from which it is a logical deduction that detection of cellulose in an organism identifies it as a plant. Or again, a student of the Protozoa is soon told of the surprising occurrence of strontium sulphate in the body of the radiolarian *Thalassicola*, from which fact it is logical to assume that the detection of strontium in a radiolarian is sufficient to assign it to the genus mentioned.

The bodies of animals contain a considerable number of complex compounds, in which connection the word 'complex' signifies compounds with large molecules containing many atoms and almost certain to exist in isomeric modifications. They include, for example, hormones, enzymes and other proteins, and nucleic acids with peptides, sugars and amino acids. The task confronting the biochemical taxonomist is to compare and contrast compounds of the same class and performing the same function in different animals, both as regards their properties and their distribution in different organs of the body.

In traditional taxonomy classification is based not on one characteristic (monothetic) but on as many as possible (polythetic). Hence it should be important for the biochemist to make his comparisons in respect of several types of compounds. This is neither a simple nor a quick operation.

No chemical conclusions can be reached until the compounds concerned have been separated and identified, and the successful accomplishment of this, where it has so far been achieved, is largely due to recent developments in two laboratory techniques, chromatography and electrophoresis.

Chromatography offers a method by which the constituents of a mixture can be separated and subsequently identified: it depends on the different rates at which compounds in solution move along a porous medium such as a piece of filter paper (paper chromatography) or a column of powdered chalk (column chromatography). Anyone can easily see this by putting a spot of coloured ink near the edge of a piece of blotting paper and then holding the paper with the edge dipping under the surface of water. This not only shows the separation of the components of the ink, it also shows how essentially simple is the principle underlying one of the most important laboratory arts of the last few decades. Virtually the only elaborations of the ink-spot experiment are changes in the solvent used and in the choice of the absorbent paper.

To understand the process in slightly greater detail, imagine that a piece of protein has been hydrolysed and thus converted into a mixture of amino acids. A drop of this mixture is placed near the edge of the paper and at other points drops of known amino acids are added as markers (figure 10). When the solvent

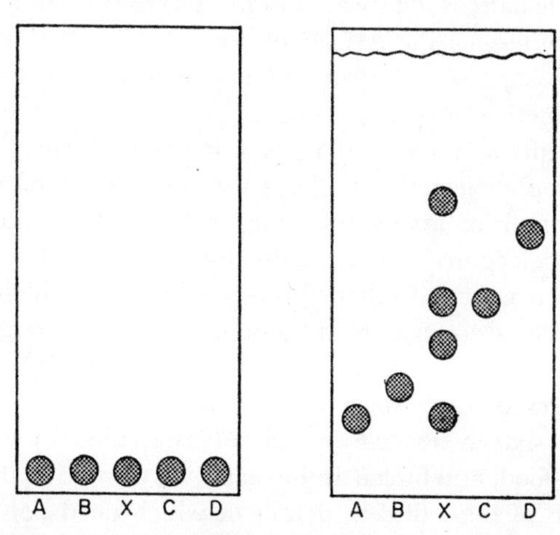

Fig. 10. Example of separation of four compounds by paper chromatography.

has completed its journey to the other edge of the paper (in practice the run is usually arranged from above downwards) the paper is dried and the spots are made visible by a spray, usually of ninhydrin. In the example illustrated, X appears to contain A and C but not B or D. It should be added that by this method quantities of a few millionths of a gram can be detected.

Electrophoresis is a similar movement of dissolved substances through a fixed medium, but in this case the movement is brought about by electrical potential difference. Just as the ions in a solution of an electrolyte move towards the oppositely charged electrodes, so will the charged molecules of compounds such as proteins. A drop of a solution of mixed proteins is put in the middle of a strip of absorbent paper immersed in a buffer solution of appropriate pH. A potential difference of a few hundred volts is provided by means of an electrode at each end, whereupon the protein molecules move, rather slowly, towards one end or the other. Known compounds are again used as markers.

Some recent results from work of this kind may be given. From investigation of the amino acids it has been shown that the insulin of pigs and whales is identical, and insulin from oxen, horses and sheep are different: the ACTH of pigs is different from that of oxen: the vasopressin of oxen contains argenine but that of the pig has lysine.

Such results as these lead us to hope that biochemistry may help in the phylogenetic problems that do not yet receive their best solution from taxonomy. One or two actual examples of biochemistry's contributions may be given.

There is a genus of ciliate Protozoa known as Blepharisma, and six species were examined by Seshachar and Saxena in 1963 to determine the different amino acids present in their cystoplasm. Five cultures of each species were used, the individuals were starved for sixteen hours to avoid contamination by excreta or undigested food, and fifteen amino acids were found in them. The distribution of these fifteen, details of which need not be given here, suggested the following phylogenetic relationship between the six species:

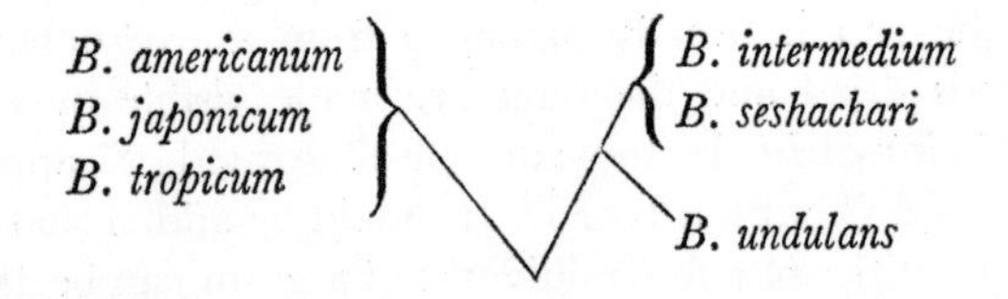

It may be added that the distribution of amino acids in the bodies of whole animals or in selected tissues has been used to distinguish species among other Protozoa, and also among mosquitoes and lampreys.

A second example, in which the result is not quite so definite, concerns four genera of Heteroptera (bugs), namely *Aspongopus*, *Dysdercus*, *Lygaeus* and *Oxycarenus*. They too were deprived of food but were allowed water to drink before they were killed. Members of each species were divided into two groups. Those in one group were analysed whole, while those in the other were dissected and the blood, muscle, fat and nerves were analysed separately. Fifteen amino acids were identified in the alcoholic extract. A diagrammatic representation of the conclusions is as given:

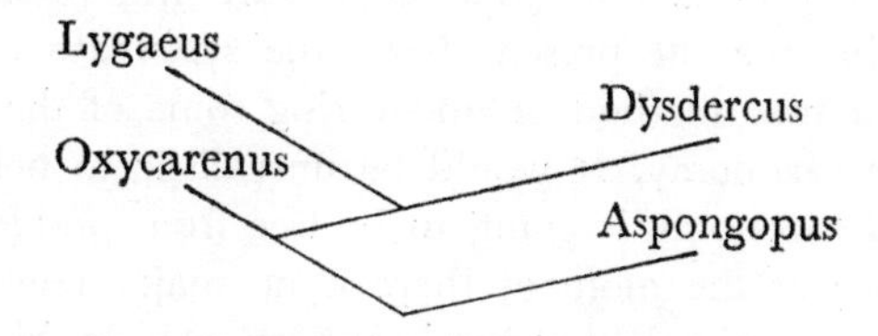

Entomologists have placed *Aspongopus* in the family Pentatomidae, *Oxycarenus* and *Lygaeus* in the Lygaeidae, and *Dysdereus* in the Pyrrhocoridae; but the biochemists would opine that *Lygaeus* resembles *Dysdercus* more closely than it does *Oxycarenus*, in respect of external appearence, while the Lygaeidae and Pentatomidae are distinguished by the presence of ocelli in Lygaeidae and their absence from Pentatomidae; so that the unbiased taxonomists are asked to decide whether the union of *Lygaeus* and *Oxycarenus* in one group on the basis of their ocelli is a better representation of their biological status than the union of *Lygaeus* and *Dysdercus* in another group on the basis of amino acids.

An extensive study of enzymes in a large number of vertebrates has been made by Kaplan, who has come to the tentative conclusion that it should be possible to classify animals on the basis of their enzymes, and further, that changes in enzyme structure may have been of significance in the establishment of new species.

There is however no information as to whether a change of enzymes was a necessity for the appearance of a new species, or whether it preceeded the change, or whether it followed the establishment of a new species, due to other causes as the species improved its adaptation to its environment.

Kaplan's suggestions lead on to the statement of H. C. Crick, who wrote that biologists should be prepared to accept a protein taxonomy based on the amino acid sequences in the proteins of an organism, and on the differences between these sequences as found in different species. He described such sequences as 'the most delicate expression possible' of the phenotype of an organism, and added that much phylogenetic information is concealed in them.

There is reason to hope that the future will resolve many of the doubts that at present beset the systematists who are pioneering a new method of uncovering some of the obscurer problems of taxonomy. It would be unrealistic to believe that biochemical taxonomy is going to be free from problems of its own making; at the moment there is no major regrouping of either plants or animals that has been instigated by biochemical generalizations. But biochemistry can contribute information which, used in combination with the evidence from other sources, may help to indicate the directions in which taxonomy should be developed.

10

Zoological Nomenclature

Every zoologist knows that the true, official, or scientific name of an animal consist of two words, and that the first is the name of its genus and the second is the name of the species. These two words are now, by international agreement, very reasonably to be known as the generic name and the specific name respectively: together they form the binomen and thus represent the long-established system of binominal nomenclature. The system has shown itself to be simpler in theory than in practice, and it is advisable that a taxonomist should be offered at least an introduction to the principles of onomatography.

First of all, the names are either Latin words or 'latinized' words, which means that they are invented or composed so as to look like and to be treated as Latin words. The remark is sometimes made that this was a deliberate choice because such words would be intelligible to men of all nations, but the truth is that the use of Latin is a vestige from the days when all intellectual work was written in that language. It follows that a little knowledge of Latin is very desirable.

In the names of animals the generic name is a noun in the nominative singular, and the specific name is most often an adjective, e.g. *Homo sapiens*, which can be translated to mean 'a wise man'. It follows that the adjective must agree with its noun in gender:

Lepus timidus (masc.) *Talpa europea* (fem.) *Distomun hepaticum* (neut.)

Many generic names, like two of those just mentioned, are the Latin (or the Greek) names given to animals by the early Roman or Greek writers, but there are not enough of these for all the genera of animals now known. Hence some names are other nouns of a more or less suitable nature, like *Helix*, a spiral, for snails and *Unio*, a pearl, for oysters; and some are adjectives pressed into use for this purpose, like *Limulus*, askew. Many of the most appropriate generic names are those which zoologists have composed, because they draw attention to some characteristic of the animals, and are therefore more easily remembered. Examples are *Stomoxys*, acid-mouth, and *Oryctolagus*, burrowing hare.

Specific names are just as diverse in their nature. In addition to the ordinary adjectives already mentioned, which are the commonest, there are many created compound adjectives, commendable because, like composed generic names, they are easily associated with some feature of the species. Names like *erythrocephalus*, red head, *trilineatus*, three-lined, and *silvicola*, living in the wood, are apt and are of infinite possibilities.

Two other types of specific name should be mentioned. Sometimes a common animal was known to the Romans by two names, and in this case both may be adopted, the second being a noun in apposition with the first. Examples are *Bos taurus*, the ox, and *Salmo salar*, the salmon. Alternatively the second noun may suggest any suitable attribute, as in *Boa constrictor* or *Orca gladiator*.

Lastly, the specific name may be a noun in the genitive case, either in the singular, like *Pieris brassicae*, or in the plural, like *Demodex folliculorum*. And the most acceptable of these are the names of zoologists, whose work in their science is thus remembered. There may be found *Nyctis daubentoni*, *Sorex granti*, *Trypanosoma brucei*, and many others. These examples introduce the convention that in zoology (but not in botany) the proper names of persons and places are not printed with a capital but with a small initial letter.

Omitting a small number of names that are foolish because they are of an absurd length or are meant to be humourous (*Marichisme* = Mary kiss me) a reference may be made to names constructed from a random selection of letters. This peculiar habit is permitted by custom, and is tending to become more frequent as names increase in number and ignorance of Latin is commoner than heretofore; but the disadvantage of names like *Oxynoe* or

Torix or *Mycelis* is that they have no meanings which would make them easier to remember.

As well as the two words of the binomen, there are certain additions to a name which may or may not be present in all circumstances. The first of these is the name of the 'author', that is of the zoologist who first gave the animal its name, and very often this is followed by the date. Thus one may read *Latimeria chalumnae* Smith, 1938. It should be noted that there is no comma between the specific and the author's names: also that these two adjuncts, though not an essential part of an animal's name, are often helpful.

An example will explain this. The butterfly's name *Polyommatus alexis* was incorrectly used for two different species by two different entomologists: addition of the author's name makes it possible to interpret *P. alexis* Rottenberg as *P. astrache* and *P. alexis* Denis as *P. icarus*, these being the true names of the specimens examined and misinterpreted.

Attention should be directed to the use of brackets or parentheses, which in zoological nomenclature have a special meaning. Too often they are wrongly put round the name of the author of a species, as if to suggest that the name is of minor importance. This may be so, but this is not the significance of the parentheses, which indicate the fact that the name of the species has undergone a change. Changes of name, which are unavoidable, will be discussed below, so that a single example is all that is necessary here.

When Linnaeus named the lion he called it *Felis leo*, so that its full name was *Felis leo* Linnaeus, 1758. Nearly two hundred years later a revision of the family Felidae caused the lion to be placed in the genus *Panthera*, and in consequence its name became *Panthera leo* (Linnaeus), 1758. The other reason for parentheses arises when a genus is divided into sub-genera; so that one of the species, as an example, is named *Ammothea* (*Achelia*) *echinata* (Hodge).

The system of nomenclature, a few of the basic essentials of which have now been mentioned, is of long-standing growth. It was originated because of the confusion that existed when writers could take it upon themselves to change the name of any animal that they were writing about, for any reason that seemed to appeal to them.

Two attempts, abortive in practice, to introduce a universal system were made by the British Association in 1842 and the International Congress of Zoology, meeting in Paris in 1889, and it was not until the third Congress met in Leyden that a committee was appointed to deal with the whole complex problem. The first truly cosmopolitan set of *Régles Internationales de la Nomenclature Zoologique* was produced in 1901. Since then the Code of Zoological Nomenclature has come under review at every meeting of the International Congress, and a completely new edition was published in 1961. It represents sixty years' experience and is a noteworthy example of international co-operation. It deserves the close attention of every zoologist.

The purpose of such a set of worldwide rules is obvious enough:

1. To ensure uniformity in the spelling and printing of names.
2. To avoid ill-constructed and otherwise objectionable names.
3. To determine the valid and legitimate name of every animal.
4. To define the circumstances in which a name may be changed.
5. To direct the procedure when a family or a genus is subdivided.

An essential feature of the *Régles Internationales* is the manifest but frustrating fact that they and their composers can do no more than suggest and recommend what have seemed to be the wisest principles. They can neither compel acceptance nor punish transgression: their chief purposes are to emphasize priority and secure continuity.

The first of these ideals imposes the seemingly simple rule that the name used in the first adequate description of an animal determines its binomen. The edible frog was named *Rana esculenta* by Linnaeus in 1758, and so it has remained ever since. But too often there has been a disadvantage in the strict application of priority, as when some earnest student, thumbing the pages of an obscure and local journal, has chanced upon a description of a familiar animal, published at an earlier date than the name in common use. Thereupon, a re-christening is called for, with the Rule of Priority as sponsor. As an example, there is some reason for believing that an early microscopist described *Amoeba proteus* under the name of *Chaos*, and for a while it was widely advocated that *Amoeba,* used and known all over the world since 1795, should be abandoned, and that *Chaos* should take its place. Fortunately such an absurdity was avoided under the pressure of common sense.

In such crises one of three things will happen.

1. The law will be enforced or followed and the change will be recognized. A long period may elapse before all zoologists become aware of the change and adopt the new name: in this interval the animal is being referred to by two names.

2. Nobody takes much notice of a piece of silly digging in unproductive lodes, and the familiar name, theoretically incorrect, continues in use. This is undesirable, as it discredits the *Régles Internationales.*

3. The International Commission is informed of the threat to established custom, and decrees that the actually later name shall continue in use. The name becomes a 'nomen conservandum'.

In the nature of things, name-changing is sometimes essential and is accepted by all. It is often resented by those who do not fully understand the course of events, which can be made plain by taking a fictitious animal through its baptismal history.

1. Abel finds a strange creature, believes it to be new and describes it under the name of *Axus rapidus* sp. nov. Later it appears as *Axus rapidus* Abel.

2. Boaz points out that the animal does not belong to *Axus* but to a related genus *Exus.* Thus it is named *Exus rapidus* (Abel), with *Axus rapidus* as a synonym.

3. Cain recalls that the specific name *rapidus* has already been used in the genus, and suggests that the species be renamed *Exus celer* (Abel). It now has two synonyma.

4. David revises the genus *Exus* and splits it into several new genera. In this process *E. celer* becomes *Ixus celer* (Abel) with three synonyma.

5. Esau revises the family and finds that the type species of *Ixus* is identical with the previously described *Oxus celer*, the type of the genus *Oxus*. Therefore all species of *Ixus* transfer to *Oxus*. But *Ixus celer* cannot become *Oxus celer* and therefore resumes its original specific name, and becomes *Oxus rapidus* (Abel). It now has four synonyma.

6. Festus revises the genus *Oxus*, but being a cautious systematist does no more than introduce sub-genera. In one of these our species now finds its place and its name becomes *Oxus* (*Uxus*) *rapidus* (Abel).

7. Gad is the first zoologist ever to see *O. rapidus* alive and to keep it alive. He discovers that it is the larval form of a rare insect discovered half a century ago and seen on only three subsequent occasions. All six names mentioned above thus become invalid.

This facet of taxonomic history, though admittedly fictitious, is in no way exaggerated. Any zoologist who has made a special study of any group could recall similar genuine instances.

At this point the distinction between invalid and illegitimate names may be explained. Invalid names are those that do not agree with the conditions laid down in the International Code; all of the seven synonyma in the above fable are invalid. A legitimate name is one that when published was accompanied by an illustrated description and a reference to a type specimen. This follows that a name may be legitimate but invalid.

The two aims of the *Regles Internationales* have obviously been harder to attain than was at first to be expected, but fewer problems exist in the naming of the higher taxa. A conspicuous feature of the Code is that it offers neither rules nor recommendations for the names of orders, classes and phyla. This is justified because the names of most of the orders and of nearly all the classes and phyla have been in use for many years and have long been acceptable as suitable and familiar. Moreover, the number of phyla and classes is not large enough to give a zoologist any difficulty in learning and remembering them.

There is invariable obedience to the rule that the names of families be made by adding *-idae* to the stem of the name of the type-genus, and of sub-families by a similar addition of *-inae*, but the International Code goes little further. It 'recommends' the termination *-oidea* for super-families and *-ini* for tribes.

The suggestion that names of orders should end in *-ida* did not secure acceptance. There is little point in uniformity for uniformity's sake, and no one would support a system that extended Worcestershire and Herefordshire to Kentshire and Sussexshire. Further, the new names may have no meaning or a stupid meaning. The word Scorpionida would imply an order of animals that were like scorpions, which is about as sensible and helpful as saying that salt tastes salty.

This chapter has provided an introduction to the nomenclatural side of zoological systematics, while neglecting many of the finer details that it includes. It may end with a word of warning. There is an ever-present risk that the art of nomenclature may acquire a disproportionate importance in the minds of zoologists themselves. This is a complex which it is not difficult to understand, and sufficient examples have been known to evoke the comment that zoology is the study of animals, not the writing of animal's names. Those whose bonnets contain this particular bee are nearly always sincere and well-meaning, but they do not promote the cause of taxonomy, and are, unkind though it may be to say so, a nuisance.

11

The Language of Taxonomy

Every occupation of man produces its own characteristic language properly known as its jargon, developed by its followers so that they can conduct its operations and discuss its problems with each other. The phrases used by a naval officer on board a man of war have but little meaning to the farmer in his fields; the words of a cricketer are not those of a printer.

Every branch of science similarly produces its own vocabulary, an inevitable accompaniment of its progress and without which progress itself would be impossible. The vocabulary quickly develops into a jargon which is almost unintelligible to followers of a different vocation, a consequence which, though unavoidable, is the origin of a criticism all too often heard of the writings of scientists. The advance of science, and indeed the advance of civilization, depends not so much on the intelligibility of science to musicians, mechanics and members of Parliament as upon the ability of scientists to understand one another.

One of the disturbing features of the language of zoology in general is the vagueness or uncertainty of the exact meaning of so many of the words that zoologists use. Scientists of all kinds have for long been accustomed to bend the meanings of words to suit themselves as an alternative to the invention of new ones. To justify so serious an accusation as this a few examples may be given.

Biology has largely grown round the idea of the cellular structure of organisms, but there is little agreement on the definition of the word 'cell'. To the question, 'Are the Protozoa unicellular or non-cellular?', either answer may be returned. Recently the present writer asked the author of a book on cells why he had not defined a cell, and the answer was that he had not dared to try.

Most animals are parasites, yet it is just as difficult to find a definition of a parasite. That there may be asked the question 'Do you consider that a cow is a parasite on a field of grass?' is almost inconceivable, yet its existence well illustrates the fog that surrounds the word. And finally when correspondence in a science-education journal sought the differences between symbiosis and commensalism, no firm conclusion was reached.

So deplorable a state of affairs must be avoided if systematics is to justify its reputation as the most scientific of all sciences. If the opinions and beliefs of taxonomists, or even the ways in which they set down their work, are to be of value to others, their words should be defined and their phrases explained in a glossary of the conventional kind. There are, however, two characteristics of their language that should be mentioned first.

One of these is the rather surprising change of meaning that has come over two important words, surprising because one of the characteristics of the language of science is that its words generally tend to preserve their meanings unchanged by the passage of time, with a constancy that is not found among the words of everyday speech. The two words referred to are 'systematics' and 'taxonomy'.

Formerly these two words were taken to be synonymous and were used with the same meaning. In fact, systematics is an anglicized form of the word 'Systema' used by Linnaeus in the title of his famous book, and 'taxonomy' is an invented piece of biological jargon with the same idea behind it. Their difference today is to be found in its right place below.

The second characteristic is a tendency to a reification of words, a weakness that is by no means confined to biologists. Many years ago T. H. Huxley wrote 'All we know about the force of gravity or any other so-called force is that it is a name for an observed order of facts'; and he thus emphasized the point that the invention of a name for a process or phenomenon seems to endow it with a new individuality and somehow to make it more real, more potent, than it was before. Chemists, for example, have to resist the temptation to write of catalysis as if to do so were to explain the hastening of a reaction: ethologists are prone to say that a kinesis is the responsible cause of an animal's behaviour. Similarly, systematists are inclined to write as if phylogeny or ontogeny or other abstract nouns were names of active factors, exerting their own particular influence on each other and on the course of evolution. But this is not so, and in more cautious writing these similar words are to be used as names for observed orders of facts.

In the glossary that follows some of our more important words are given fuller linguistic treatment than is customary in zoological books. It is hoped that some readers will appreciate this, that others will be led to do so, and that all will be willing to condone it. In any case, zoology cannot suffer if zoologists are given a chance to read some description of the words that they are compelled to use.

I

Systematics 1650. (i) Formerly a synonym of 'taxonomy', q.v. (ii) Latterly, the 'scientific study of the kinds and diversities of organisms and of any or all relationships among them'. (Simpson)

It should be noted that in spite of the final *s*, systematics is a singular noun. The Latin suffix *-icus* implied possession or association, as in *aquaticus*, belonging to water, and its more usual English form is *-ic* for adjectives, as in 'toxic', while nouns, like hysterics, ballistics, cybernetics and mathematics, lose the *u* and retain the *s*.

Taxonomy. (Gk. tassein, to arrange; nemein, to deal out) 1813.
The study of the rules, principles and practice of classifying, particularly of organisms.

Phylogeny. (Gk. phulon, a race; genesis, descent) 1870. The study of the evolution of groups or of species in the animal or plant kingdoms.

Classification. (Lat. classis, a class; facere, to make.) The arranging of animals or plants in groups, based on their resemblances and differences.

Nomenclature. (Lat. nomen, a name; calare, to call.) The making and giving of distinguishing names to each and every group in the classification of animals or plants.

Onomatography. (Gk. onomos, a name; graphein, to write) 1964. The practice of the correct writing of the names of animals or plants; analogous to bibliography, the correct writing of the names of books.

II

Hierarchy. A tradition of mediaeval times divided the inhabitants of Heaven into three hierarchies, each of which was subdivided into three orders:

First Hierarchy	*Second Hierarchy*	*Third Hierarchy*
Order 1 Seraphim	Order 4 Dominions	Order 7 Principalities
Order 2 Cherubim	Order 5 Virtues	Order 8 Archangels
Order 3 Thrones	Order 6 Powers	Order 9 Angels

Thus hierarchy (Gk. hieros, sacred; archein, to govern) literally sacred ruling, has come to mean organization into grades or ranks, and its similarity to zoological classification cannot be missed. The correct adjective is hierarchic, not hierarchical or hierarchal.*

Phylum. (Gk. phulon, a race.) Introduced into taxonomy by Ernest Haeckel in 1877.

Class. (Lat. classis a division of the Roman people.) Also an army or fleet; also any division, in general terms.

Cohort. (Gk. chortos, a feeding place; Lat. cohors, an enclosure.) Hence, by transference, a tenth part of a Roman legion, and thence a throng or multitude.

Order. (Lat. ordo, a straight line, as in *ordo vitium*, a row of vines.) Whence a rank of soldiers and, later, a political group.

Family. (Lat. familia.) Introduced into taxonomy by Butschli in 1780.

* Incidentally, a most interesting opuscule could be written on the influence of classical scholarship on scientific thought.

Tribe. (Lat. tribus, a tribe.) Originally a third part of the Roman people; whence any group of people ethnologically related.

Genus. (Lat. genus, generis, birth or breed.) Whence, by transference, a race or stock.

Species. (Lat. specio, look at, behold.)

III

A considerable number of words have been invented by taxonomists at various times to describe the many phenomena that they have encountered, and it is thought that definitions of a selection of these may be helpful if included here.

Agamospecies. A species that reproduces asexually.

Allopatric. Occurring apart, in regions separated by a natural barrier.

Archetype. A pattern or model on which the design of other objects is based. The adjective is archetypal, not archetypic.

Binomial. Any expression consisting of two terms, not necessarily related. Obsolescent in biological nomenclature.

Binominal. Having two names, as, in biology, the name of the genus and the name of the species.

Biospecies. A population consisting of individuals able to breed together.

Character. The overall nature of an object; the sum of its several characteristics.

Characteristic. A feature peculiar to one object or to one group of related objects.

Cladistic. Based on community of lines of descent or evolution.

Cladogram. Tree-like diagram, illustrating the sequence of the evolutionary descent of any group of organisms.

Definition. A precise description of an organism (or other object) in terms of the characteristics by which it can be recognized.

Dendogram. Any branching, tree-like diagram, illustrating the relationships of any kind between different objects.

Diagnosis. A short statement of the characteristics that distinguish organisms of one taxon from those of another.

Monothetic. Based on one or only a few characteristics; e.g. a classification of plants based on the number of stamens.

Palaeospecies. A group of related, extinct, organisms, which may be assumed to have been able to interbreed.

Phenogram. A tree-like diagram, illustrating the results of numerical taxonomy.

Polythetic. Based on many characteristics, most but not necessarily all of which are shown by every member of a group.

Sympatric. Occurring together in the same region.

Taxometrics. Quantitative or numerical estimates of similarities between organisms or groups of organisms, leading to their ordering into higher taxa.

Taxon. Any unit in taxonomy, such as an order, family or genus.

12

Systema Animalium

The only logical way in which to end this book will be to offer a suggested form of animal classification, in which the principles outlined in earlier chapters are invoked to produce a system that attempts to combine the practical usefulness of the Linnaean hierarchy with the theoretical phylogeny of the present day.

Ideally, this would take the form of a considerable volume, in which one can visualize the roll-call of phyla, classes and orders in their correct places, printed on the right-hand pages, with their authors' names and dates included. The left-hand pages would be reserved for explanatory notes, dealing with alternative names, obsolete names and references to the literature of taxonomy. So fascinating a task as this is beyond the scope of one chapter.

In the following pages diagnoses are given for the phyla and higher groups, but not for classes. The taxa named do not descend below the class or sub-class, save in the three instances of the important classes of birds, insects and mammals.

KINGDOM ANIMALIA

Sub-kingdom Protozoa. Non-cellular animals.

Phylum Protozoa. Animals of small size, whose bodies are not divisible into separate cells. Usually one nucleus is present, though sometimes two occur.

Sub-kingdom Parazoa. Cellular animals in which the cells, arranged in two or three layers round a central cavity, function independently.

Phylum Porifera. Sponges, in which the cells are of the collared flagellate type. The colony is supported by a skeleton of spongin, with or without spicules of chalk or silica. The principal opening is exhalent.

Class Calcarea

Class Hexactinellida

Class Demospongiae

Sub-kingdom Metazoa. Cellular animals in which the cells act in co-operation

Grade Radiata. Metazoa in which the body exhibits radial symmetry, and has reached the epithelial level of organization.

Sub-grade Coelenterata. Aquatic animals, usually sessile when adult, in which two layers of cells surround a coelenteron. Cnidoblasts, thread-cells, are the characteristic weapons.

Phylum Cnidaria. Coelenterata in which hydroid and medusoid phases usually occur in alternation.

Class Hydrozoa

Class Scyphozoa

Class Anthozoa

Phylum Cnidaria. Marine coelenterates, free-swimming by means of rows of fused cilia. Nematocysts absent.

Class Tentacula

Class Nuda

Grade Bilateria. Metazoa in which the body is bilaterally symmetrical, and anterior and posterior ends are distinct.

Sub-grade Acoelomata. Metazoa in which the region between the alimentary canal and the body wall is filled with undifferentiated cells or mesenchyme.

Phylum Platyhelminthes. Free-living and parasitic forms in which the alimentary canal has a single opening, osmoregulatory organs take the form of solenocytes, and the reproductive organs are often hermaphrodite.

Class Turbellaria

Class Trematoda

Class Cestoda

Sub-grade Pseudocoelomata. Metazoa in which the body cavity is formed by large vacuoles in the cells outside the alimentary canal.

Phylum Aschelminthes. Pseudocoelomata with no trace of segmentation. No cilia. Nervous system usually a collar.

Class Nemertini
Class Nematoda
Class Rotifera
Class Chaetognatha

Sub-grade Coelomata. Metazoa in which a coelomic cavity often round the alimentary canal is lined with mesodermal epithelia.

Infra-grade Protostomia. Coelomata in which the blastopore of the embryo becomes the mouth of the adult.

Phylum Phoronida. Marine colonial zooids, surrounded by a membranous tube. Excretory organs are nephromyxia.

Phylum Polyzoa. Small colonial zooids, with ciliated tentacles and flame bulbs as excretory organs.

Class Phylactolaemata
Class Gymnolaemata

Phylum Brachiopoda. Unsegmented, marine coelomata, covered by a bivalve shell in dorsal and ventral positions. There are coelomoducts but no nephridia.

Class Inarticulata
Class Articulata

Phylum Mollusca. Coelomata, usually apparently unsegmented, in which the body is protected by a calcareous shell and is divided into head, foot and visceral hump.

Class Aplacophora
Class Polyplacophora
Class Monoplacophora
Class Gasteropoda
Sub-class Prosobranchia
Sub-class Opisthobranchia
Sub-class Pulmonata

Class Scaphopoda
Class Lamellibranchiata
Class Cephalopoda

Phylum Annelida. Metamerically segmented worms, normally with setae on most segments and with a cuticle of collagen. Excretory organs are nephridia.

Class Polychaeta
Class Oligochaeta
Class Hirudinea
Class Archiannelida

Phylum Onychophora. Caterpillar-like arthropods, with many pairs of unjointed legs. Excretory organs are coelomoducts.

Class Onychophora

Phylum Arthropoda. Metamerically segmented animals in which the body cavity is haemocoelic and there are neither cilia nor nephridia. Formation of a head is manifest and each somite normally carries a pair of jointed appendages, at least one of which serve as jaws.

Sub-phylum Pycnogonida. Marine arthropods in which the head carries a proboscis and three pairs of appendages. There are normally four pairs of legs: the abdomen is reduced to a small sac.

Class Pantopoda

Sub-phylum Chelicerata. Arthropoda in which the body is divided into two parts, the prosoma with six pairs of appendages, including chelate chelicerae.

Class Merostomata
Sub-class Xiphosura
Class Arachnida
Sub-class Latigastra
Sub-class Caulogastra
Class Tardigrada
Class Pentastomida

Sub-phylum Mandibulata. Arthropoda in which the head carries a pair of antennae and a pair of mandibles.

Class Crustacea
- Sub-class Branchiopoda
- Sub-class Ostracoda
- Sub-class Copepoda
- Sub-class Mystacocarida
- Sub-class Branchiura
- Sub-class Cirripedia
- Sub-class Malacostraca

Class Diplopoda
Class Chilopoda
Class Pauropoda
Class Symphyla
Class Insecta
- Sub-class Apterygota
  - Order Thysanura — Order Collembola
- Sub-class Pterygota
- Infra-class Palaeoptera
  - Order Ephemeroptera — Order Odonata
- Infra-class Polyneuroptera
  - Order Dictyoptera — Order Isoptera
  - Order Plecoptera — Order Chelevtoptera
  - Order Orthoptera — Order Dermaptera
- Infra-class Oligoneuroptera
  - Order Coleoptera — Order Megaloptera
  - Order Raphidioptera — Order Planipennia
  - Order Mecoptera — Order Trichoptera
  - Order Lepidoptera — Order Diptera
  - Order Siphonaptera — Order Hymenoptera
  - Order Strepsiptera
- Infra-class Paraneuroptera
  - Order Psocoptera — Order Mallophaga
  - Order Anopleura — Order Thysanoptera
  - Order Homoptera — Order Heteroptera

Infra-grade Deuterostomia. Coelomata in which the blastopore becomes the anus of the adult.

Phylum Pogonophora. Benthic animals, their long thin

bodies enclosed in a tube of chitin, and with a mass of tentacles at one end.

Phylum Echinodermata. Marine animals with a calcareous skeleton in the mesoderm. Larvae bilaterally symmetrical, adults radially symmetrical, usually with five rays. A water-vascular system, which operates the tube-feet, opens at a madreporite.

Class Stelleroidea
Sub-class Asteroidea
Sub-class Ophiuroidea
Class Echinoidea
Class Holothuroidea
Class Crinoidea

Phylum Chordata. Coelmata with a notochord at some period of development, similarly with pharyngeal gill clefts; with a hollow dorsal nerve cord and a post-anal tail.

Sub-phylum Acrania. Chordata with no skull
Class Hemichordata
Class Urochordata
Class Cephalochordata

Sub-phylum Craniata. Chordata in which the brain is enclosed in a skull.

Infra-phylum Agnatha. Craniata in which the skull bears no jaws.
Class Cyclostomata

Infra-phylum Gnathostomata. Craniata in which jaws are present.

Super-class Anamniota. Craniata in which there is no amnion round the embryo.
Class Chondrichthyes
Sub-class Selachii
Sub-class Bradyodonti
Class Osteichthyes
Sub-class Crossopterygii
Sub-class Actinopterygii
Class Amphibia

Super-class Amniota. Craniata in which an amnion surrounds the embryo.
- Class Reptilia
- Class Aves
  - Super-order Impennae
  - Super-order Neognathae
- Class Mammalia
  - Sub-class Prototheria
    - Order Monotremata
  - Sub-class Metatheria
    - Order Marsupialia
  - Sub-class Eutheria
    - Infra-class Unguiculata
      - Order Insectivora
      - Order Edentata
      - Order Cheiroptera
      - Order Primates
    - Infra-class Glires
      - Order Rodentia
      - Order Lagomorpha
    - Infra-class Mutica
      - Order Cetacea
    - Infra-class Ferrungulata
      - Order Sirenia
      - Order Pinnipedia
      - Order Artiodactyla
      - Order Carnivora
      - Order Proboscoidea
      - Order Perissodactyla

13

Last Thoughts

The attempt to expound the principles of taxonomy which has filled the foregoing chapters should lead in conclusion to a review of the science as a whole. If its strength and its weaknesses are thus brought more closely together a juster impression is more likely to follow. That taxonomy has made great progress cannot be denied, but there remains for appraisement the question of how far taxonomists have succeeded, and are succeeding, in achieving their aims.

The function, indeed the expressed purpose of taxonomy is to discover, to define, and to elaborate the methods that may best be used to produce an acceptable classification of the animal kingdom. The implication of this function is that taxonomy is to be regarded as a science in its own right, with fundamental concepts, some at least of which ought to be applicable to the classification of objects of any kind, and should not be confined to the organisms with which biological taxonomy alone deals. Complementary to this there follows the realization that classification is an art, to be admitted not to the category of fine arts but to the more practical one of useful arts.

For the basic truth, which was almost the first concept to be emphasized in Chapter 1, is that we classify because we must. Faced as we now are by many more than a million different kinds of animals, we must classify before our mutual conversations become intelligible among ourselves. The first observations that we make are to the effect that all these animals vary at different stages in their lives and in different parts of the world.

At first sight these facts seem to be an encouragement rather than the reverse. The great number of animals assures us opportunities to exercise both skill and judgement. If there were but a score or so of animals known to us, their arrangement in suitable groups would not be a difficult problem. If the animals vary considerably, as obviously they do, again the art of classification should be simplified, for classifications must depend on differences and distinctions. It would be quite impossible to classify a hundred newly minted florins.

Thus inspired, the taxonomist approaches his task, and almost immediately there is borne in upon him the unwelcome truth that permeates the whole of taxonomy, *ab ovo usque ad mala*. This is no more than the fact that the animals of the world were not meant to be classified. Whatever may be the purpose of cosmic life on this planet, there is no evidence of intent that its animals should be grouped or graded or sorted to meet any human needs.

The human taxonomist therefore attempts to classify objects that are not inherently amenable to classification, and his efforts may be compared to attempts to translate Newton's *Principia* from its original Latin into English limericks. The initial material is simply not suited to the process. But the attempt has to be made, and was indeed made from the earliest days of intellectual effort, to continue with varying intensity ever since. It is no less prevalent today and it will indoubtedly continue, always exposing the fundamental difficulty.

Few aspects of pure science illustrate better than does taxonomy the truth that 'a little knowledge is a dangerous thing'. The operation of classifying appeared to be so simple in its beginnings; for obviously a horse was different from a donkey but at the same time it was more like a donkey than a cow. And so on, throughout all the range of familiar and conspicuous animals. With the added assumption that all horses, all donkeys, all cows were sufficiently alike to be regarded as constant entities there arose the idea of distinctive species, followed by that of different categories of similarity.

Trouble appeared and grew as the number of known species multiplied at an increasing rate. We have seen how these difficulties were met in different ways by different zoologists, whose varying opinions could only stimulate a search to discover the 'best' system of classification. The most generally accepted answer to this problem was the extraordinary one that the best classification was a natural one.

This was extraordinary because classification itself is essentially artificial, a search for regularity among organisms which exist in an untidy profusion and showing variation wherever this adapts them to survive in variable circumstances. From this impasse taxonomy was offered escape by the suggestion that the doctrine of evolution should point the way, imposing upon the system of classification the responsibility or duty of displaying in summary form the evolutionary history of the animal kingdom.

Here was a counsel of perfection that introduced almost as many difficulties as it sought to remove. Evolution in any one group of animals occurs in different places at different rates and at different times, orders in one class may be found to be more closely allied than those in another class of the same phylum, and the families in some orders may be in the same way different from those in another. To express such facts as these, as well as others less obvious but equally significant, in a list of named taxa printed normally on paper in two dimensions is equally found to be impossible. Many authors have tried to improve on this by diagrammatic representation of a three-dimensional nature, and on a small scale such devices have offered an improved interpretation of phylogeny. But they are not easy to draw, and perhaps they need more than three dimensions, at which point the ordinary zoologist is inclined or compelled to abandon his efforts.

Taxonomists have done their best with their eight obligate taxa and with the sub-divisions that so noticeably increased their precision, and the most dramatic consequence has been the admission of reasonable doubt as to the validity of 'species'. Almost from the beginning of biology the concept of the definite, individual, isolated species, had seemed to be the one unquestioned axiom of systematics. Claiming the equivalent of divine establishment, accepted without question from Aristotle to Cuvier, the species had to surrender its crown in the 'new' systematics.

Taxonomists, feeling that their bedrock was crumbling beneath their feet, made gallant efforts to preserve the species concept from disappearance. A species might have to be replaced by a population, but the population was said to consist of a selection of variously qualified species, some of which have been discussed in an earlier chapter.

Such general discontent leads sooner or later to a recasting of ideas and a revision of method. For example, the number of characteristics that were compared in the process of making the taxa at any grade in the hierarchy was increased, measurement and enumeration were emphasized, and the computers were called in to determine the significance of the facts. Or alternatively the biochemist was enlisted to analyse the body's components and thus to supply data free from prejudice. These later devices are at present promising, but have not yet reached the stage of adolescence.

There were others who looked at the matter from their own points of view, among whom may be mentioned the philosophers. They discoursed learnedly, as is their wont, about inductive and deductive methods, about apriorism and empiricism, and similar topics. No doubt they brought to taxonomy a logical stability and an aura of academic respectability; but it is doubtful whether the practical systematist benefited much from their deliberations.

A more hopeful approach was made by the geneticists. They could argue, with no risk of contradiction, that all similarities and also all differences between animals must be attributable to their genetic constitutions, and that most problems must yield to adequate knowledge of genotypes. We await no more than a complete mapping of the nucleotide sequences in all the deoxyribonucleic acids in all the chromosomes in all the animals. If this thesis be accepted, the admission must follow, if indeed it was ever doubted, that all lines of demarcation between taxa are drawn on wholly arbitrary judgements. The spreading of the concept of the polytypic species emphatically supports this, and most of the causes of disagreement and sources of discussion tend to vanish.

There are some who feel that avoidable difficulties are introduced by acceptance of the idea that only one system of classification has claims to serious consideration. All orthodox classification is based on structural characteristics, the significance of an animal's construction being assumed to be of paramount importance. There has never been a reasoned attempt to justify this claim of morphology to its position of supremacy; almost the only modification has been the attempt to add the features of the internal organs to those externally visible.

Can it be that there was a greater possibility hidden in Oken's system than his contemporaries and successors have realized? Lorenz Oken (1779–1851) was a philosopher–biologist who laid emphasis on the responses of the central nervous system to the impulses received from the sense organs. He suggested a classification of animals into five groups:

1. Dermatozoa, dependent chiefly on the sense of touch.
2. Glossozoa, dependent chiefly on the sense of taste.
3. Rhinozoa, dependent chiefly on the sense of smell.
4. Otozoa, dependent chiefly on the sense of hearing.
5. Ophthalmozoa, dependent chiefly on the sense of sight.

This is very close to the ordinary arrangement of animals into invertebrates, fishes, amphibians, reptiles, birds and mammals respectively, but its chief feature is that it represents a new approach to the problem of classification. Such revolutionary methods are often worth more consideration than they receive.

Reference has been made to the suggestion that two systems of classification should be recognized, one for the scientist who is willing to include phylogeny and another for the naturalists and amateurs. The suggestion has never been implemented, but it deserves consideration. In justification of such support as the idea, it may be pointed out that our own species is satisfactorily subjected to different classifications for different purposes. It is divided into males and females on many occasions, into tall and short for recruitment to the police force, old and young for pension rights, married and single for taxation, clever and stupid for educational grants, and black and white for playing cricket in South Africa. The human race is less diverse than the rest of the animal kingdom, and if a comparable view were to be taken of the animal world a similiar multiplicity of classifications might be created.

There is no *a priori* reason why such a scheme should not be operated: we have long been accustomed to speak of birds, beasts and fishes, and of omnivora, carnivora and herbivora, and the pundits have dismissed such divisions as being valueless. But in fact the physiologist is not really concerned with the fact that a mole and a camel have seven cervical vertebrae, the ecologist is unmoved by the presence or absence of a coelom, the ethologist watches his chosen animals behaving as they are wont, but he does not stop to count the number of setae on their femora or tarsi. There is reason to believe that the production of a system of systems might well justify the time and thought devoted to it. Taxonomists have travelled a long road, but they would be the first to admit that the end is not yet on the near side of the horizon.

A scientist would not wish to end his book on a note of pessimism, but there is an irresistible temptation to recall a quatrain whose author is unknown to me and who cannot therefore be personally thanked for writing:

> 'They told him it couldn't be done:
> Smiling, he went right to it
> And tackled the thing that couldn't be done;
> And couldn't do it.'

Bibliography

ALSTON, R. E. and TURNER, B. L. 1963. *Biochemical Systematics.* Prentice Hall.

CAIN, A. J. 1966. *Animal Species and their Evolution.* Hutchinson.

CALMAN, W. T. 1949. *The Classification of Animals.* Methuen.

HUXLEY, J. S. 1940. *The New Systematics.* O.U.P.

ROTHSCHILD, LORD 1968. *A Classification of Living Animals.* Longmans.

SAVORY, T. H. 1962. *Naming the Living World.* E.U.P.

SIMPSON, G. G. 1945. *The Principles of Classification and a Classification of Mammals.* Bull. Amer. Mus. Nat. Hist. lxxxv. pp. 1–350.

SIMPSON, G. G. 1961. *The Principles of Animal Taxonomy* Columbia University Press.

SOKAL, R. S. and SNEATH, P. H. A. 1963. *Principles of Numerical Taxonomy.* Freeman.

Systematic Zoology. passim.

Index